# A Photographic Atlas for the Microbiology Laboratory

**Michael J. Leboffe**
*San Diego City College*

**Burton E. Pierce**
*San Diego City College*

**Morton Publishing Company**
925 W. Kenyon, Unit 12
Englewood, Colorado 80110

Printed in the United States of America

10  9  8  7  6  5  4  3

ISBN:  0-89582-308-X

# Preface

*A Photographic Atlas for the Microbiology Laboratory* is intended to act as a supplement to introductory microbiology laboratory manuals. It is not designed to replace them, nor is it intended to replace actual performance of the techniques. Rather, the photographs are supplied to help in interpretation of results.

The *Atlas* is divided into nine sections, each devoted to a particular component of a typical introductory microbiology course. Topics within the first eight sections typically include the following components.

**Purpose**  The purpose describes *why* the procedure is done. If used to detect the presence of a particular enzyme, it is listed in this section.

**Medical Applications**  Medical applications are described for most procedures as a specific application of the *purpose*. In some cases, a procedure is diagnostic of a particular group. More often, a procedure is used in combination with many others during identification of a pathogen. In these instances, some important pathogens with their diseases and their result for the test are listed.

**Principle**  The principle comprises several features. First, each procedure is used to probe some aspect of bacterial physiology, morphology or biochemistry. These are addressed and the procedure is put into the context of the living microbe. Chemical structures of relevant metabolic intermediates have been supplied. Second, the theory behind the procedure itself is discussed. Chemical structures of reagents are usually included. Third, while instructions on *how* to perform each procedure are omitted (since these may be obtained in your laboratory manual), helpful hints to avoid common pitfalls are provided.

The ninth section is devoted to a survey of viruses, protozoans and fungi that might be seen or discussed in an introductory microbiology class. Medical comments are also included where relevant.

Also included are several appendices. The first three provide detail on glycolysis, Krebs Cycle and fermentation. A list of commercial microscope slides used in preparation of this *Atlas* is supplied for instructors who wish to match slides available in their lab with the content of this book. Lastly, a listing of selected references is provided for students who wish to expand on the information provided herein.

Microbiology lab can be a fun and rewarding experience. We hope you find it to be so, and that the *Atlas* contributes to your success.

Michael J. Leboffe
Burton E. Pierce
September 1995

# Acknowledgments

No project of this type can be completed without assistance from colleagues. We were overwhelmed at the willingness of so many people to participate and are grateful to the many individuals for their contributions. Thanks to all of you.

Dianne Anderson — San Diego City College
David Brady — San Diego City College
Joyce Costello — San Diego City College
Deborah Durand — UCSD Department of Medicine
Jeff Foti — Ward's Natural Science Establishment, Inc.
Susan Garrison — Carolina Biological Supply
Deborah Gelfand — San Diego City College
Randall Kottel — Mesa College
Bill Morse — Ward's Natural Science Establishment, Inc.
Darla Newman — Mesa College
Michele Pierce — San Diego City College
Brett Ruston — San Diego City College
Rachel D. Schrier — UCSD Department of Pathology
David Singer — San Diego City College
Students of Summer 1995 General Microbiology Class, San Diego City College
Joy Sussman — Becton Dickinson Company
Robert-Eli Anthony Wheeler and Margaret Wheeler
Clayton Wiley — University of Pittsburgh
Gary Wisehart — San Diego City College
Alan Zeglarski — Instrument Service Co.

Special thanks are warranted for certain individuals. Robert Waddell, Thomas Lee, Larry Kuritzky and William McClellan of Scientific Instrument Company, Temecula, CA made an Olympus BX40 Photomicroscope with automatic exposure available to us. Without their assistance, the photomicrographs would have been impossible. We are also grateful to Jerome Hunter, President, Carol Dexheimer, Business Manager, and Richard Massa, Vice President of Instruction of San Diego City College for their assistance in arranging use of San Diego Community College District facilities through the Civic Center Program.

The advice and support of Doug Morton and Christine Morton of Morton Publishers were greatly appreciated. Their confidence and enthusiasm in this project made it possible for us to focus on creativity rather than on more mundane matters. Thanks also to Joanne Saliger of Ash Street Typecrafters, Inc. for her contributions in designing the *Atlas*.

Lastly, thanks and love go to our wives Karen Leboffe and Michele Pierce for their emotional support and understanding that our temporary preoccupation with bacteriological media and photographs does not diminish our love for them.

# Table of Contents

# Bacterial Growth Patterns

## BACTERIAL COLONY MORPHOLOGY

**Purpose** Recognizing different bacterial colony morphologies on agar plates is useful for distinguishing between different species in a mixed culture. Once identified as different, cells from a colony may be transferred to a sterile medium to begin a pure culture.

**Principle** When a single bacterial cell is deposited on an agar surface which supplies its nutrient needs, it begins to divide. One cell makes two, two make four, four make eight . . . one billion make two billion, and so on. Eventually, a visible mass of cells is found on the plate where the original cell was deposited. This mass of cells is called a *colony*.

Figures 1-1 through 1-14 show some of the variety of bacterial colony forms and characteristics. The basic categories include colony shape, margin (edge), elevation, color,

and texture. Colony shape may be described as *circular*, *irregular*, or *punctiform* (tiny). The margin may be *entire* (smooth, with no irregularities), *undulate* (wavy), *lobate* (lobed), *filamentous*, or *rhizoid* (branched like roots). Colony elevations include *flat*, *raised*, *convex*, *pulvinate* (very convex), and *umbonate* (raised in the center). Colony texture may be *moist*, *mucoid*, or *dry*. Color is self-explanatory, but may be combined with optical properties such as *opaque*, *translucent*, *shiny* or *dull*.

You should not be overwhelmed by these terms. Most of these terms replace descriptive phrases, so they are intended to make colony morphology description easier, not more difficult. Nor should you feel there is always a single, correct description for each organism on every medium. Rather, use these terms in a way that is meaningful to you and realize it's more important to recognize each when you see it than to memorize all the descriptive terms for a particular species.

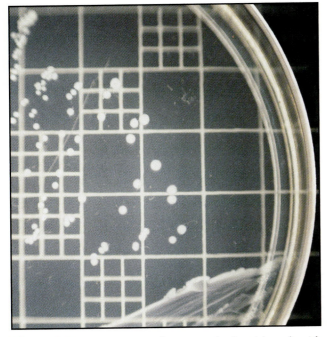

**Figure 1-1.** *Enterococcus faecium* colonies (viewed with transmitted light) are cream colored and circular with an entire margin.

**Figure 1-2.** White, raised, circular and entire colonies of *Staphylococcus epidermidis* viewed with reflected light.

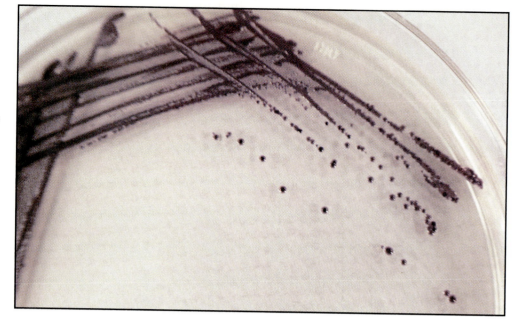

**Figure 1-3.** The purple colonies of *Chromobacterium violaceum*.

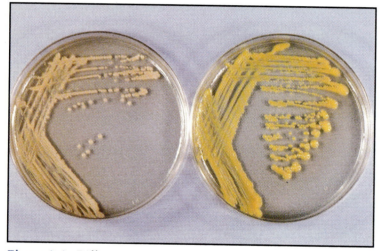

**Figure 1-4.** Difference in pigmentation between closely related *Micrococcus roseus* (left) and *M. luteus* (right).

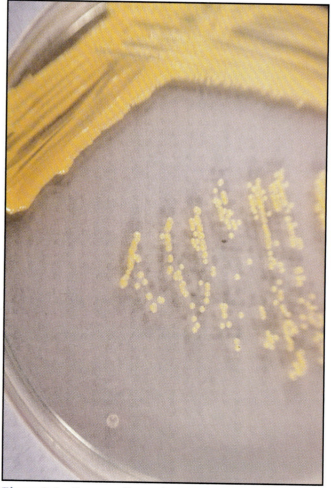

**Figure 1-6.** Convex yellow colonies of *Micrococcus luteus*.

**Figure 1-5.** Pigment production may be influenced by environmental factors. *Serratia marcescens* produces an orange pigment grown at 25°C (left) and does not at 37°C (right).

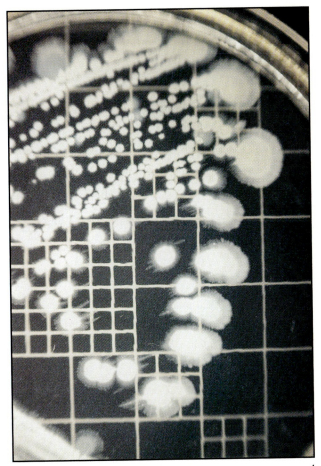

**Figure 1-7.** *Bacillus subtilis* with an opaque center and spreading edge.

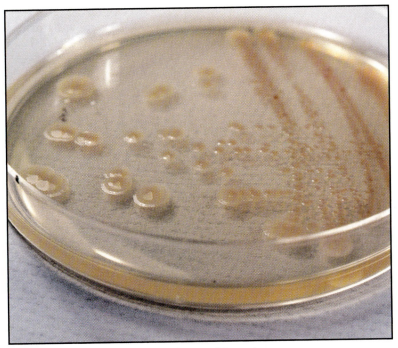

**Figure 1-8.** Shiny, umbonate colonies of *Serratia marcescens*.

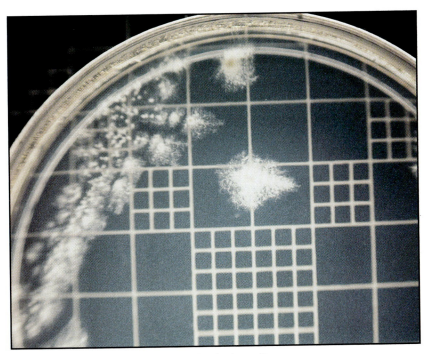

**Figure 1-9.** Irregular rhizoid growth of *Clostridium sporogenes*.

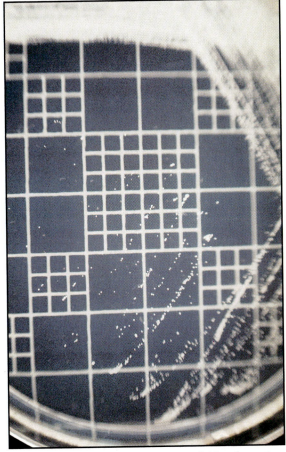

**Figure 1-10.** Punctiform colonies of *Mycobacterium smegmatis*.

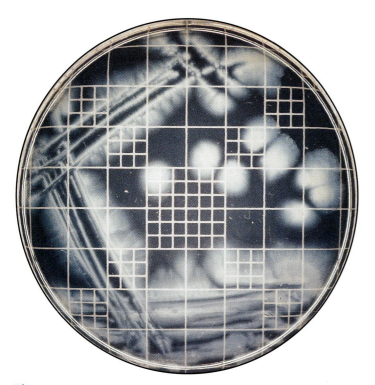

**Figure 1-11.** *Alcaligenes faecalis* demonstrates spreading and translucent growth.

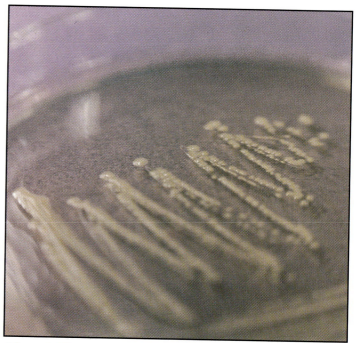

**Figure 1-12.** Raised colonies of *Klebsiella pneumoniae*.

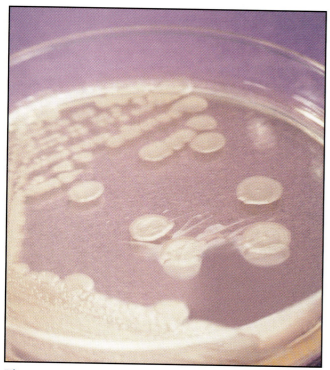

**Figure 1-13.** *Bacillus subtilis* colonies illustrating a raised margin and dull surface.

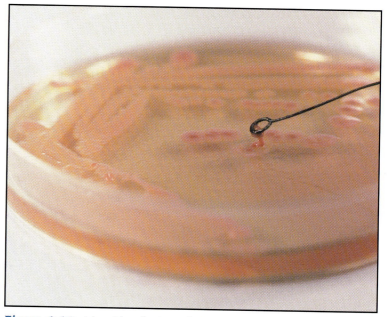

**Figure 1-14.** Mucoid colonies of *Klebsiella pneumoniae* on deoxycholate medium.

*A Photographic Atlas for the Microbiology Laboratory*

# GROWTH PATTERNS IN BROTH

**Purpose**   Characteristic growth patterns in broth may be of use in identifying an unknown bacterium.

**Principle**   Growth pattern in broth may be used to distinguish between microbes during identification. Each may be the result of specific cell structures. Growth patterns may be *uniform fine turbidity* (cloudiness), *flocculent* (clumps of growth), or *sediment*, as shown in Figure 1-15. Figure 1-16 shows two types of surface growth: a *ring* and a *pellicle* (surface membrane).

**Figure 1-15.** Growth patterns in broth. From left to right: *Bacillus cereus* (sediment), *Clostridium sporogenes* (flocculent), and *Pseudomonas aeruginosa* (uniform fine turbidity).

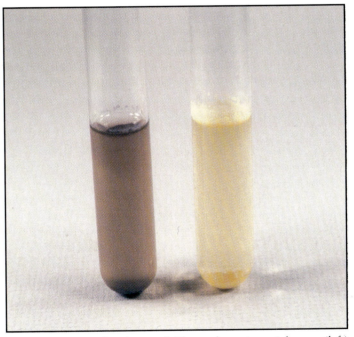

**Figure 1-16.** Broth cultures of *Chromobacterium violaceum* (left) and *Corynebacterium xerosis* (right) illustrating a surface ring and a pellicle, respectively. Both also exhibit fine turbidity and a slight sediment.

# AEROTOLERANCE—AGAR DEEP STAB TUBES

**Purpose** Recognizing the aerotolerance of a specimen is of use in identifying an unknown bacterium.

**Principle** Bacteria are classified according to their aerotolerance—that is, their ability to grow in the presence of oxygen. Organisms requiring oxygen are *strict (obligate) aerobes*, whereas those that cannot live in the presence of oxygen are *strict (obligate) anaerobes*. Between these two extremes are *facultative anaerobes* that grow better aerobically than anaerobically, and *aerotolerant anaerobes* that grow the same under both conditions. *Microaerophiles* require oxygen, but at a concentration less than is found in the atmosphere.

Agar deep tubes may be used to determine aerotolerance. A stab inoculation is made to the bottom of the agar in a test tube (see Fig. 1-17). With such a small surface area for diffusion of oxygen into the agar, the aerobic zone extends to a depth of only about 1 cm from the surface. After incubation, the portion(s) of the tube supporting growth allows aerotolerance determination. Growth only at the top indicates an aerobe. Growth only in the lower portion indicates an anaerobe. Uniform growth throughout indicates an aerotolerant anaerobe, whereas growth throughout but more in the aerobic zone indicates a facultative anaerobe. Microaerophiles grow in the aerobic zone, but below the surface.

**Figure 1-17.** Agar deep stab tubes indicating microbial aerotolerance. From left to right: *Clostridium sporogenes* (strict anaerobe), *Staphylococcus aureus* (facultative anaerobe), and *Pseudomonas aeruginosa* (strict aerobe).

# ANAEROBIC CULTURE METHODS

**Purpose**   These procedures allow growth of anaerobes under anaerobic conditions.

**Principle**   Bacteria typically are grown under aerobic conditions which support growth of all aerotolerance categories except for strict anaerobes. Creating conditions for anaerobic growth requires extra effort, and many systems have been developed. Two anaerobic culture methods are discussed in this section.

One anaerobic culture medium is *thioglycolate broth*. It contains sodium thioglycolate, a reducing agent that reduces free $O_2$ to water, thus making the broth anaerobic. A dye, such as methylene blue or resazurin, may be included to indicate where oxygen remains in the medium. Resazurin dye is pink if oxidized and colorless if reduced; methlyene blue is blue if oxidized, colorless if reduced. Depending on which dye is used, thioglycolate tubes will have a pink or blue band near the surface where oxygen diffuses in. In addition to allowing growth of anaerobes, thioglycolate may also be used to determine aerotolerance. Interpretation of aerotolerance is the same as for agar deep tubes (see Fig. 1-18).

A second method of growing anaerobes is the anaerobic jar (see Fig. 1-19). A gas generator packet is opened, water is added, and the lid is immediately clamped on the jar. Sodium borohydride and sodium bicarbonate in the packet react with the water and produce $H_2$ and $CO_2$ gases. Palladium then catalyzes the conversion of $H_2$ and $O_2$ to water, as shown in Figure 1-20.

Removal of free oxygen produces anaerobic conditions within the jar and the entire jar may then be incubated at the appropriate temperature. A methylene blue strip is included to indicate if anaerobic conditions are actually created. Methylene blue is blue if oxidized and colorless if reduced. Therefore, a functional anaerobic jar should have a white strip.

Aerotolerance may be determined by inoculating plates with the same organisms and incubating one aerobically and the other anaerobically in the jar (see Figs. 1-21 and 1-22).

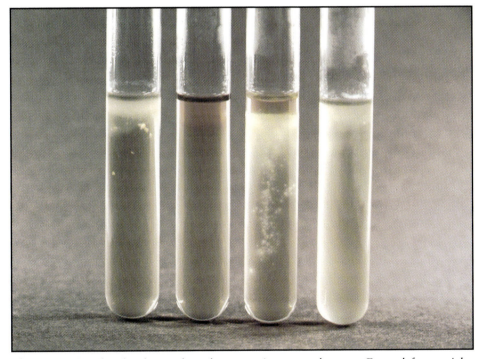

**Figure 1-18.** Thioglycolate tubes demonstrating aerotolerance. From left to right: *Moraxella catarrhalis* (strict aerobe), uninoculated control, *Clostridium sporogenes* (strict anaerobe), and *Providencia stuartii* (facultative anaerobe). Note the pink bands in the control and *C. sporogenes* tubes due to $O_2$ oxidizing resazurin.

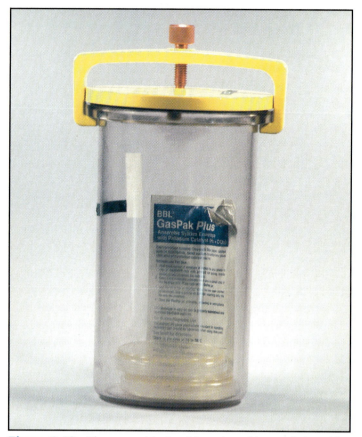

**Figure 1-19.** The anaerobic jar. Note the white methylene blue strip and the open packet which has discharged $H_2$ and $CO_2$ gases. The palladium catalyst is contained in the packet.

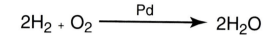

$$2H_2 + O_2 \xrightarrow{\text{Pd}} 2H_2O$$

**Figure 1-20.** Conversion of $H_2$ and $O_2$ to water is catalyzed by palladium.

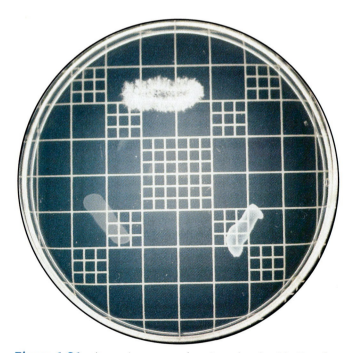

**Figure 1-21.** A nutrient agar plate inoculated with *Pseudomonas aeruginosa* (left), *Clostridium sporogenes* (top) and *Staphylococcus aureus* (right) and incubated in an anaerobic jar. Compare the amount of growth with the plate in Figure 1-22.

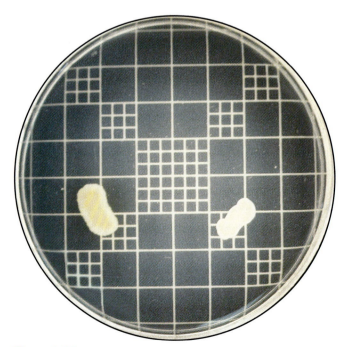

**Figure 1-22.** A nutrient agar plate inoculated with *Pseudomonas aeruginosa* (left), *Clostridium sporogenes* (top) and *Staphylococcus aureus* (right) and incubated aerobically. Compare the amount of growth with the plate in Figure 1-21.

# Isolation Techniques and Selective Media

## STREAK PLATE METHOD OF ISOLATION

**Purpose**    This procedure allows a microbiologist to obtain pure cultures from a mixed culture of microbes.

**Medical Application**    Growth of a pure culture is necessary before any tests may be run to identify a suspected pathogen.

**Principle**    Bacteria from a mixed culture are streaked over the agar surface in a pattern that deposits them further and further apart. Towards the end of the pattern, the resulting colonies should be separate from all others and may be used (*picked*) to start pure cultures.

While it is generally safe to assume a pure culture has been started from picking an isolated colony, such is not always the case. Contaminants may be slow-growing and are not yet visible on the plate. Contaminants may also be trapped in slime or among chains of bacteria. Lastly, if a selective medium has been streaked for isolation, a contaminant may be alive but not dividing (a *bacteriostatic* state) until transferred to the nonselective medium used for pure culture. Since the consequence of not producing a pure culture is almost certainly misidentification of the microbe, culture purity should be confirmed by periodic streak plates and Gram stains.

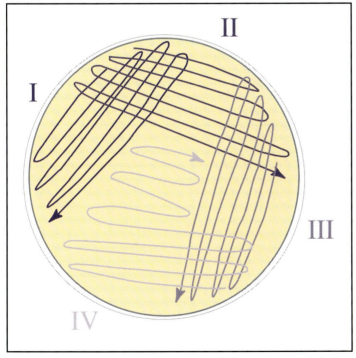

**Figure 2-1.**  The quadrant method. The agar surface is streaked as in I. After flaming the loop, the plate is rotated 90° and streaked as in II. The process is repeated for streaks III and IV.

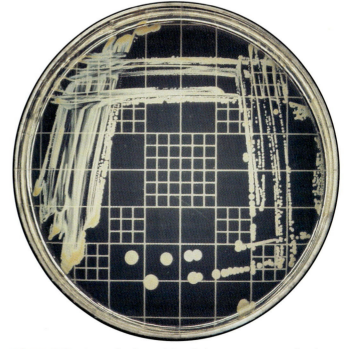

**Figure 2-2.**  A streak plate of *Staphylococcus aureus* after incubation. Note the decreasing density of growth in the four streak patterns. On this plate, isolation is first obtained in the third streak. Cells from different colonies may be transferred to a sterile medium to start pure cultures of each.

# DEOXYCHOLATE AGAR

**Purpose**    Deoxycholate agar is an undefined selective and differential medium used for isolation and differentiation among the Enterobacteriaceae—the enteric (gut) bacteria.

**Medical Applications**    Enteric bacteria are facultatively anaerobic Gram-negative rods. They may be divided into those that produce acid from lactose fermentation (the *coliforms*) and those that don't. Coliforms, including *Escherichia coli* and *Enterobacter aerogenes,* typically are nonpathogenic. The nonlactose fermenter group includes such pathogens as *Salmonella typhi* (enteric fever) and *Shigella dysenteriae* (bacillary dysentery). Deoxycholate agar allows a quick preliminary indication of whether a specimen contains enteric pathogens.

**Principle**    Deoxycholate agar contains nutrients, including lactose, sodium deoxycholate, citrate, and neutral red. Deoxycholate is a component of bile and inhibits growth of Gram-positive organisms. Citrate is included to increase the action of deoxycholate. Acid end products from lactose fermentation lower the pH (see Fig. 2-3). When the pH gets below 6.8, the colorless neutral red turns a reddish color. Thus, Gram-negative lactose fermenters will appear some shade of red, whereas Gram-negative lactose nonfermenters will remain colorless (see Fig. 2-4).

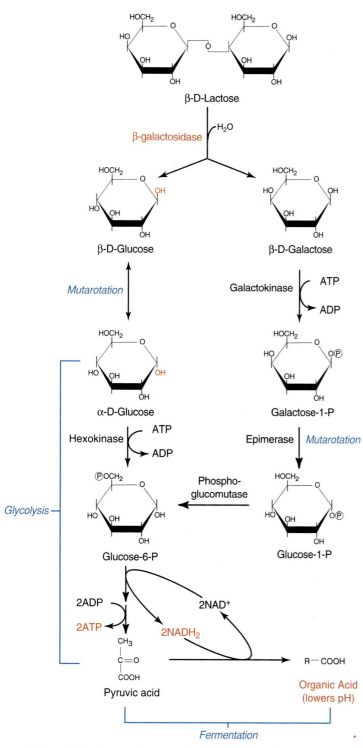

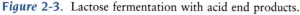

**Figure 2-3.** Lactose fermentation with acid end products.

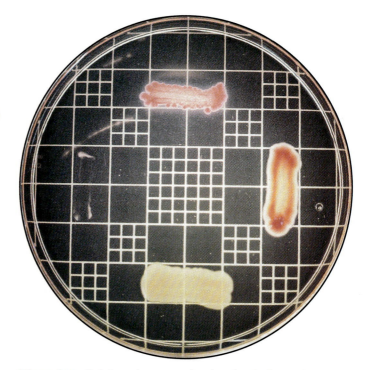

**Figure 2-4.** DOC medium inoculated with (clockwise from top): *Escherichia coli, Enterobacter aerogenes, Proteus vulgaris,* and *Micrococcus luteus.* Note the inhibition of the Gram-positive *M. luteus.* Note also the red coloring of the lactose fermenting *E. coli* and *E. aerogenes.*

# EOSIN METHYLENE BLUE (EMB) AGAR

**Purpose**   Eosin methylene blue agar is a selective and differential medium used for isolation and differentiation among members of the Enterobacteriaceae—the enteric (gut) bacteria.

**Medical Applications**   EMB agar, like DOC, allows a quick preliminary indication of whether a specimen contains enteric pathogens. Please see the medical applications for DOC medium.

**Principle**   Eosin methylene blue agar is an undefined and selective medium that contains the aniline dyes methylene blue and eosin which inhibit Gram-positive bacteria, thus favoring growth of Gram-negative enterics. In addition to the dyes, the medium also contains lactose. Lactose makes EMB a differential medium in that it allows distinction between lactose fermenters and lactose nonfermenters. Large amounts of acid from lactose fermentation (see Fig. 2-3) cause the dyes to precipitate on the colony surface, producing a characteristic green metallic sheen. Smaller amounts of acid production results in a pink coloration of the growth. Nonfermenting enterics do not produce acid so their colonies remain colorless or take on the coloration of the medium (see Figs. 2-5 and 2-6).

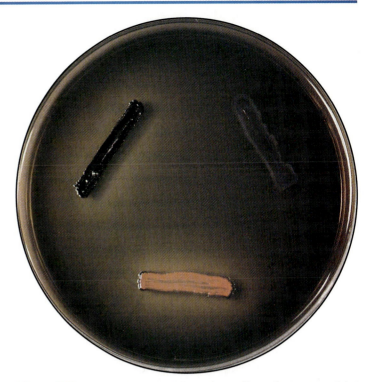

**Figure 2-5.** *Escherichia coli* (left), *Salmonella typhimurium* (right) and *Enterobacter aerogenes* (bottom) on EMB. Note the characteristic dark coloration of *E. coli* and the pink coloration of *E. aerogenes*. The difference in color is due to degree of acid production.

**Figure 2-6.** The characteristic green metallic sheen of *E. coli*.

# MacCONKEY AGAR

**Purpose**    MacConkey agar is a selective and differential medium used to isolate and differentiate between members of the Enterobacteriaceae—the enteric (gut) bacteria.

**Medical Applications**    MacConkey agar, like DOC and EMB, allows a quick preliminary indication of whether a specimen contains enteric pathogens. Please see the medical applications for DOC medium.

**Principle**    MacConkey agar is an undefined medium containing nutrients, including lactose, as well as bile salts, neutral red and crystal violet. Bile salts and crystal violet inhibit growth of Gram-positive bacteria, making MacConkey agar a selective medium. Neutral red is a pH indicator that is colorless above a pH of 6.8 and red at a pH less than 6.8. Acid accumulating from lactose fermentation (see Fig. 2-3) turns the colorless neutral red a red color. Coliforms are therefore a shade of red, whereas lactose nonfermenters remain colorless (see Fig. 2-7).

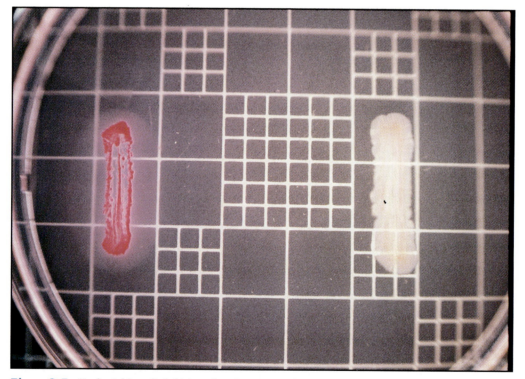

**Figure 2-7.** *Escherichia coli* (left) is red, indicating acid products from lactose fermentation. *Providencia stuartii* (right) is a lactose nonfermenter and remains colorless on MacConkey agar.

# MANNITOL SALT AGAR

**Purpose**    Mannitol salt agar is both selective and differential. It favors organisms capable of tolerating high sodium chloride concentration and also distinguishes bacteria based on their ability to ferment mannitol.

**Medical Applications**    Mannitol salt agar is used to differentiate pathogenic *Staphylococcus* species (which ferment mannitol) from the less pathogenic members of the genus *Micrococcus* (which do not). Of these, *S. aureus* is the most notable pathogen, causing, among other things, food poisoning, pneumonia, osteomyelitis, and toxic shock syndrome.

**Principle**    Mannitol salt agar is formulated with 7.5% NaCl. This makes it highly selective since most bacteria can not tolerate this level of salinity. Organisms not suited for this environment will show stunted growth or no growth at all.

Phenol red is the pH indicator included in the medium. Phenol red is yellow below pH 6.8, red at pH 7.4 to 8.4, and pink at 8.4 and above. The development of yellow halos around colonies is an indication that mannitol has been fermented with the production of acid (see Fig. 2-8). No color change or formation of pink color is a negative result (see Fig. 2-9).

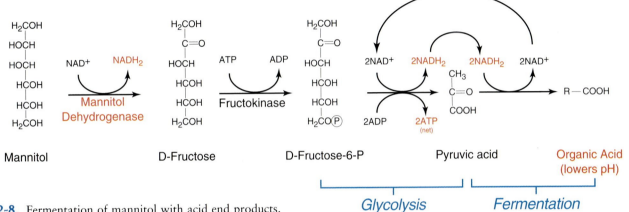

**Figure 2-8.** Fermentation of mannitol with acid end products.

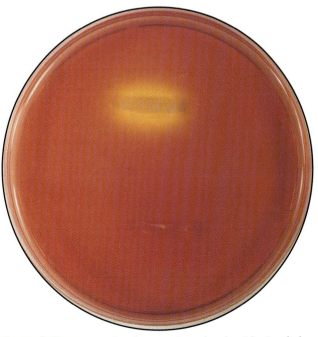

**Figure 2-9.** Mannitol salt agar inoculated with *Staphylococcus aureus* (above) and *S. epidermidis* (below). The yellow halo around *S. aureus* is due to mannitol fermentation with acid end products.

# PHENYLETHYL ALCOHOL (PEA) AGAR

**Purpose**    This medium is used to selectively inhibit growth of Gram negative organisms.

**Medical Application**    Phenylethyl alcohol agar is frequently used as a plated medium in mixture with sheep or rabbit blood to isolate hemolytic Gram-positive aerobes or anaerobes.

**Principle**    Phenylethyl alcohol agar is a selective medium which inhibits or prevents growth of most Gram-negative organisms by altering the lipid structure of the membrane and interfering with protein synthesis. The mechanism and effectiveness, however, differ slightly from species to species.

Phenylethyl alcohol agar contains nutrients, sodium chloride, and 0.025% phenylethyl alcohol (PEA). Although high concentrations of PEA can be toxic to both Gram-positive as well as Gram-negative bacteria, a concentration this low inhibits Gram-negative organisms while allowing Gram-positive bacteria to thrive.

When a mixture of Gram-negative and Gram-positive organisms are streaked onto a plate of phenylethyl alcohol agar, the Gram-positive organisms will grow and the Gram-negative organisms will show either stunted growth or no growth at all.

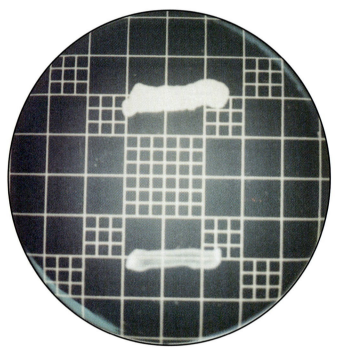

**Figure 2-10.** Nutrient agar with *Enterococcus faecium* (below) and *Escherichia coli* (above). Compare with Figure 2-11.

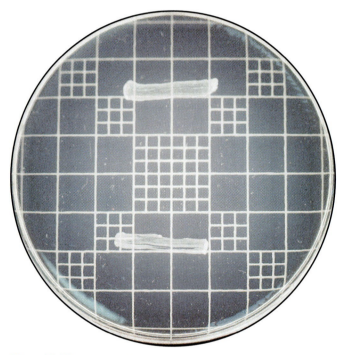

**Figure 2-11.** Phenylethyl alcohol agar with *E. faecium* (below) and *E. coli* (above). Growth of the Gram-negative *E. coli* has been retarded compared to its growth on nutrient agar (see Fig. 2-10).

# XYLOSE LYSINE DESOXYCHOLATE (XLD) AGAR

**Purpose**   Xylose lysine desoxycholate agar is a selective and differential medium used to isolate and identify enteric pathogens.

**Medical Applications**   Xylose lysine desoxycholate agar is used in identification of enteric pathogens, especially *Shigella*, from fecal samples. Members of this genus produce bacillary dysentery.

**Principle**   Xylose lysine desoxycholate agar is a selective medium due to the presence of bile salts that inhibit Gram-positive organisms. Since this medium is often used after another selective and enrichment medium has favored Gram-negative growth (such as Gram-negative broth or Hektoen enteric agar), the bile salt concentration is relatively low.

XLD is also a differential medium. It allows determination of carbohydrate fermentation as well as the ability to reduce sulfur (see Fig. 2-12).

Lactose, sucrose, and xylose are the available carbohydrates and phenol red is the pH indicator. If the pH drops below 6.8, phenol red turns yellow so, a yellow color is evidence of fermentation with acid production. In isolating and identifying enterics, acid production from lactose is of primary importance. Possible false positives may occur in lactose nonfermenting enterics that ferment one of the other carbohydrates.

The amino acid lysine is also an ingredient in this medium. Its decarboxylation produces ammonia and raises the pH. Some organisms (*e.g. Salmonella*) may initially ferment and turn the medium yellow, but then decarboxylate the lysine and turn the medium back to red.

Sodium thiosulfate is the sulfur source. Reduction of sodium thiosulfate to $H_2S$ is detected by ferric ion in the medium which produces the black precipitate FeS. Species of the enteric genera *Salmonella* and *Proteus* usually are positive for sulfur reduction.

*Figure 2-12.* XLD agar inoculated with (clockwise from top): *Proteus mirabilis* (positive for sulfur reduction), *Salmonella typhimurium* (atypically negative for sulfur reduction), and *Escherichia coli* (positive for lactose fermentation).

# Bacterial Cellular Morphology and Simple Stains

---

## NEGATIVE STAIN

**Purpose** A negative stain may be used to determine morphology and cellular arrangement in bacteria that are too delicate to withstand heat-fixing. Where accurate size determination is crucial, a negative stain may be used since there is no cell shrinkage due to heat-fixing.

**Medical Applications** Spirochaetes, such as *Treponema pallidum* (the causative agent of syphilis) are delicate cells that are easily distorted by heat-fixing. These may be observed using the negative stain.

**Principle** Stains are solutions consisting of a *solvent*, usually water or an alcohol, and a colored solute called the *chromophore*. Stains with an acidic chromophore are used for the negative staining technique. (An acidic chromophore gives up a hydrogen ion [$H^+$] which leaves it with a negative charge.) The negative charge on the bacterial surface repels the acidic chromophore, so the cell remains unstained against a colored background (see Figs. 3-1 and 3-2). Examples of acidic stains are eosin and nigrosin.

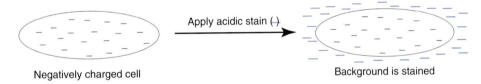

Apply acidic stain (-)

Negatively charged cell

Background is stained

**Figure 3-1.** Acidic stains have a negatively charged chromophore that is repelled by negatively charged cells. Thus, the background is colored and the cell remains transparent.

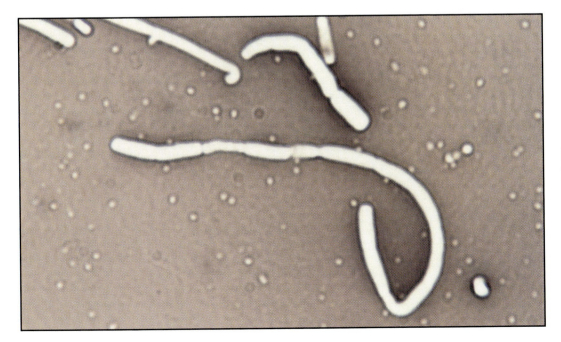

**Figure 3-2.** *Bacillus cereus* negatively stained with nigrosin (1320X).

# SIMPLE STAIN

**Purpose**    Since cytoplasm is transparent, cells are usually stained with a colored dye to make them more visible under the microscope. Cell morphology, size, and arrangement may then be determined.

**Medical Applications**    Simple stains may be used to determine cell morphology, size and arrangement. However, in a medical laboratory, these are usually determined with a Gram stain.

**Principle**    Stains are solutions consisting of a solvent and a colored solute called the *chromophore*. Since bacterial cells typically have a negative charge on their surface, they are most easily colored by basic stains with a positively charged chromophore (see Fig. 3-3). (A basic chromophore gives up a hydroxyide ion [OH⁻] or picks up a hydrogen ion [H⁺], either of which leaves it with a positive charge.) Common basic stains include methylene blue, crystal violet and safranin. Examples of these may be seen in the Gallery of Bacterial Cells section.

Basic stains are applied to bacterial smears that have been heat-fixed. Heat-fixing kills the bacteria, makes them adhere to the slide, and coagulates cytoplasmic proteins to make them more visible.

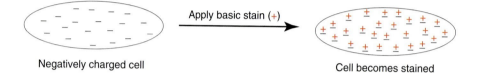

Negatively charged cell          Apply basic stain (+)          Cell becomes stained

*Figure 3-3.* Basic stains have a positively charged chromophore which forms an ionic bond with the negatively charged bacterial cell, thus colorizing the cell.

# A GALLERY OF BACTERIAL CELL DIVERSITY

## Bacterial Cell Shapes

Bacterial cells are much smaller than eukaryotic cells (see Fig. 3-4) and come in a variety of shapes and arrangements. Determining cell shape is an important first step in identifying a bacterial species.

Cells may be spheres (*cocci*, sing. *coccus*), rods (*bacilli*, sing. *bacillus*) or spirals (*spirilla*, sing. *spirillum*). Variations on these shapes include slightly curved rods (*vibrios*) and flexible spirals (*spirochaetes*). Examples of cell shapes are shown in Figures 3-5 through 3-10.

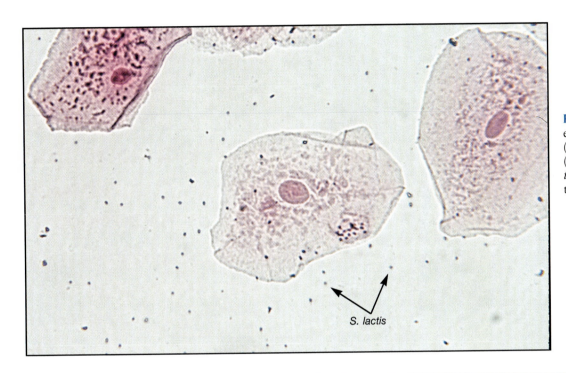

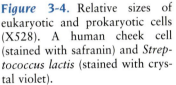

*Figure 3-4.* Relative sizes of eukaryotic and prokaryotic cells (X528). A human cheek cell (stained with safranin) and *Streptococcus lactis* (stained with crystal violet).

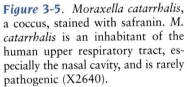

*Figure 3-5.* *Moraxella catarrhalis*, a coccus, stained with safranin. *M. catarrhalis* is an inhabitant of the human upper respiratory tract, especially the nasal cavity, and is rarely pathogenic (X2640).

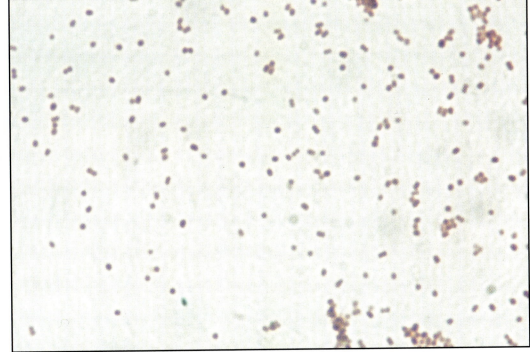

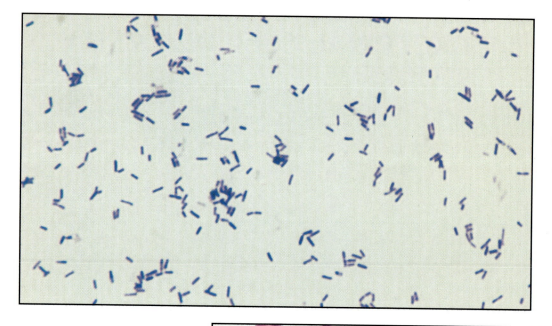

Figure 3-6. *Bacillus subtilis,* a bacillus found in soil, stained with crystal violet (X2640).

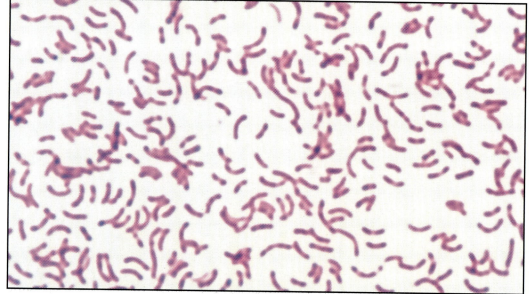

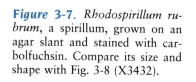

Figure 3-7. *Rhodospirillum rubrum,* a spirillum, grown on an agar slant and stained with carbolfuchsin. Compare its size and shape with Fig. 3-8 (X3432).

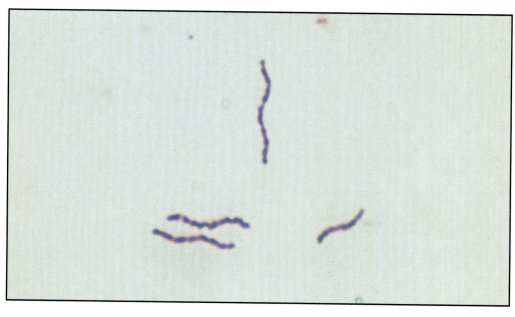

Figure 3-8. *Rhodospirillum rubrum* grown in nutrient broth and stained with carbolfuchsin. Compare its size and shape with Fig. 3-7 (X3432).

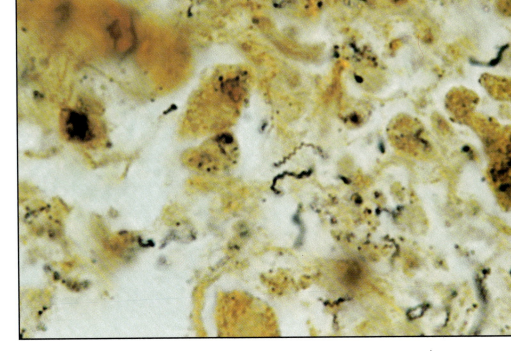

**Figure 3-9.** *Treponema pallidum,* a spirochaete, is the causative agent of syphilis in humans (X2640).

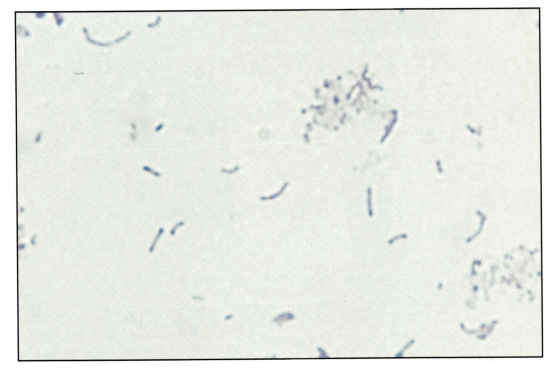

**Figure 3-10.** *Vibrio fischerii,* a luminescent vibrio found in marine habitats (X3432).

## Bacterial Cell Arrangements

Cell arrangement, determined by the number of planes in which cell division occurs and whether the cells separate after division, is also useful in bacterial identification. Cell arrangements and the prefix or root word used to indicate each are listed below.

| | |
|---|---|
| *diplo-* | pairs of cells |
| *strepto-* | chains of cells |
| sarcina | group of eight cells |
| *staphylo-* | irregular cluster of cells |

Cocci may exhibit any of these arrangements, depending on their division planes as shown in Figure 3-11. Bacilli are found as single cells or in chains. Spirilla are rarely seen as anything other than single cells. Figures 3-12 through 3-18 illustrate common cell arrangements.

Figures 3-19 and 3-20 show *Corynebacterium diphtheriae* which illustrates exceptions to the more common categories. First, the cells are pleomorphic rods, which means there is a variety of cell shapes–slender, ellipsoidal or ovoid rods–in a given sample. Second, snapping division produces either a *palisade* or angular arrangement of cells.

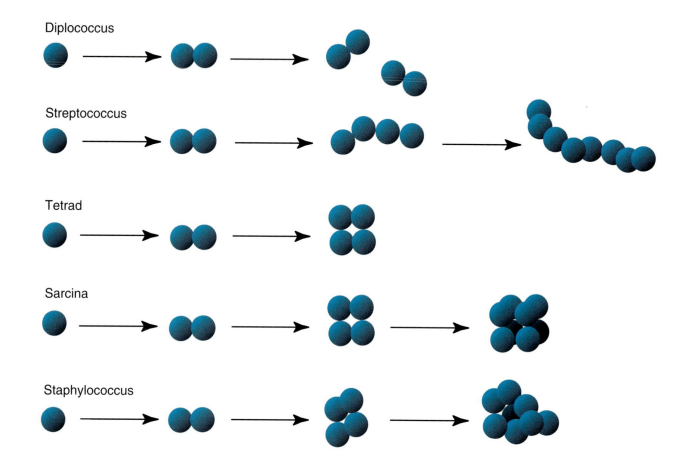

Diplococcus

Streptococcus

Tetrad

Sarcina

Staphylococcus

*Figure 3-11.* Division patterns among cocci. In diplococci, there is a single division plane and cells are generally found in pairs. Streptococci also have a single division plane, but the cells remain attached to form chains of variable length. If there are two division planes (X and Y), the cells form tetrads. Sarcinae divide in three planes (X, Y and Z) to produce a regular cuboidal arrangement of cells. Staphylococci divide in more than three planes (X, Y, Z and oblique) to produce a characteristic grapelike cluster of cells. NOTE: Rarely will a sample be composed of just one arrangement. The more complex the arrangement, the more likely scattered examples of simpler arrangements will be found. Look for the most common arrangement.

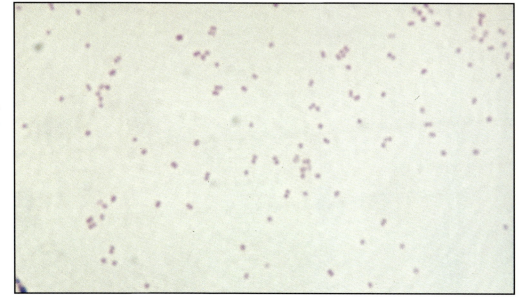

**Figure 3-12.** *Moraxella catarrhalis* is often found as a diplococcus (X2640).

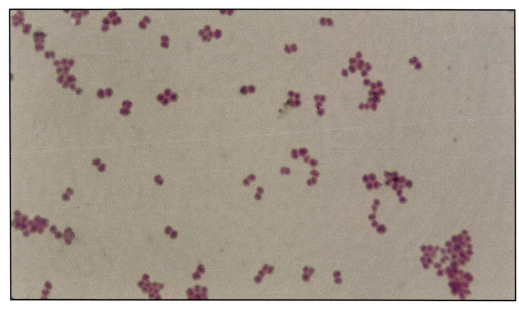

**Figure 3-13.** *Neisseria gonorrhoeae* is a diplococcus that causes gonorrhea in humans. Members of this genus produce diplococci with adjacent sides flattened (X2640).

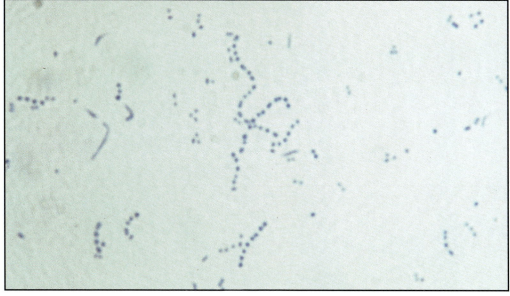

**Figure 3-14.** *Enterococcus faecium* is a streptococcus. It was stained with methylene blue (X2640).

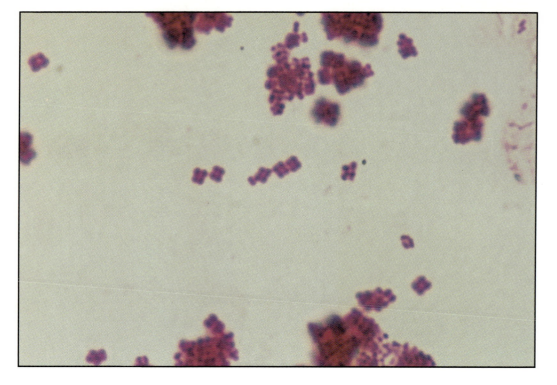

**Figure 3-15.** Tetrads of *Micrococcus roseus* stained with carbolfuchsin (X2640).

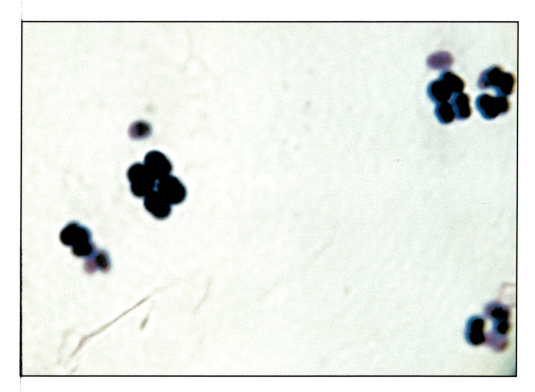

**Figure 3-16.** *Sarcina maxima* exhibits the sarcina organization (X1716).

*A Photographic Atlas for the Microbiology Laboratory*

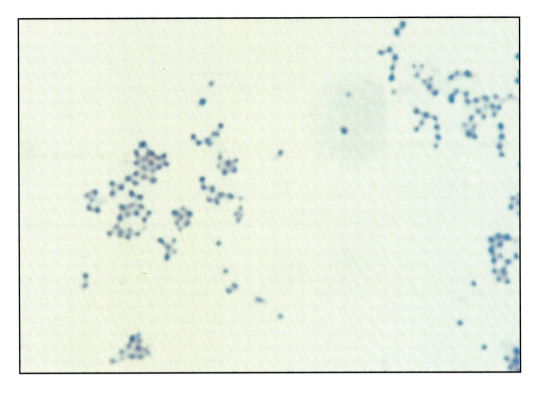

**Figure 3-17.** *Staphylococcus aureus* from broth culture showing the staphylococcus arrangement of cells. *S. aureus* is a common opportunistic pathogen of humans (X3432).

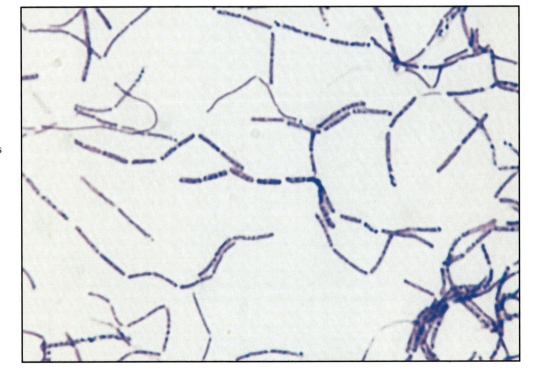

**Figure 3-18.** *Bacillus megaterium* is a streptobacillus (X2400).

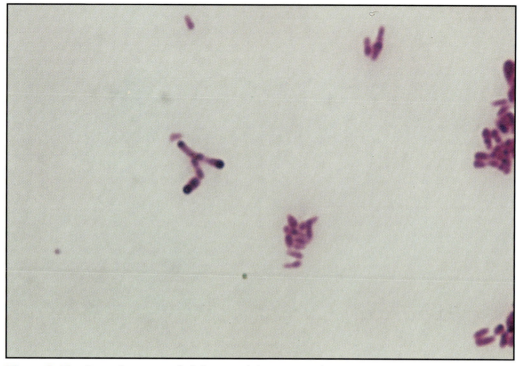

**Figure 3-19.** *Corynebacterium diphtheriae* exhibits an angular arrangement of cells produced by snapping division typical of the genus (X3432).

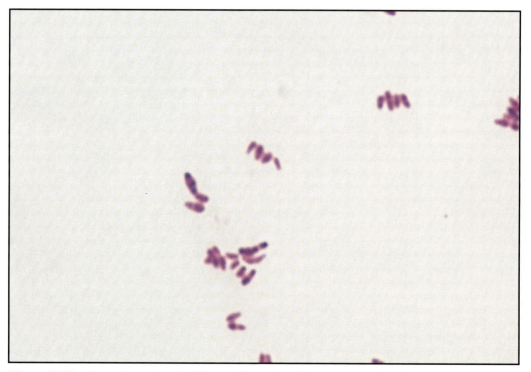

**Figure 3-20.** *Corynebacterium diphtheriae* illustrating characteristic palisade arrangement of cells (X3432).

# Bacterial Cellular Structures and Differential Stains

---

## GRAM STAIN

---

**Purpose**    The Gram stain is used to distinguish between Gram-positive and Gram-negative cells. It is probably the most important and widely used microbiological differential stain.

**Medical Applications**    The Gram stain is the first differential test run on a specimen brought into the laboratory for identification.

**Principle**    The Gram stain is a differential stain in which a decolorization step occurs between the application of two basic stains. As shown in Figure 4-1, the primary stain is crystal violet. Iodine is added as a mordant to enhance crystal violet staining by forming a crystal violet-iodine complex. Alcohol decolorization follows and is the most critical step in the procedure. Gram-negative cells are decolorized by the alcohol (95% ethanol) whereas Gram-positive cells are not. Gram-negative cells are thus able to be colorized by the counterstain safranin. Upon successful completion of a Gram stain, Gram-positive cells appear purple and Gram-negative cells appear red (see Fig. 4-2).

Electron microscopy and other evidence have allowed microbiologists to determine that the ability to resist alcohol decolorization or not is based on the different wall constructions of Gram-positive and Gram-negative cells. Gram-negative cell walls have a higher lipid content than Gram-positive cell walls. It is thought that the alcohol extracts the lipid, making the Gram-negative wall more porous and incapable of retaining the crystal violet-iodine complex, thus decolorizing it. Alcohol dehydration of Gram-positive cell walls also makes them *less* porous, so the crystal violet-iodine complex is trapped.

The decolorization step is the most crucial and most likely source of Gram-stain inconsistency. It is possible to *over-decolorize* by leaving the alcohol on too long and get red Gram-*positive* cells. It also is possible to *under-decolorize* and produce purple Gram-*negative* cells. Neither of these situations changes the actual Gram reaction for the organism being stained. Rather, they are false results due to poor technique on the part of the person performing the Gram stain. Until correct results are consistently obtained, controls of Gram-positive and Gram-negative organisms along with the organism in question may be used to eliminate concern about poor technique (see Fig. 4-3).

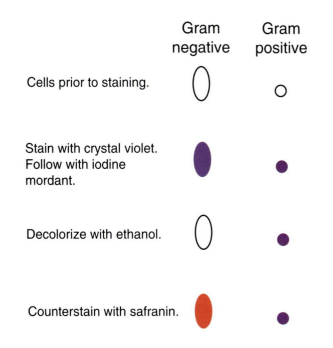

**Figure 4-1.** The Gram stain. Gram-positive cells are violet; Gram-negative are red.

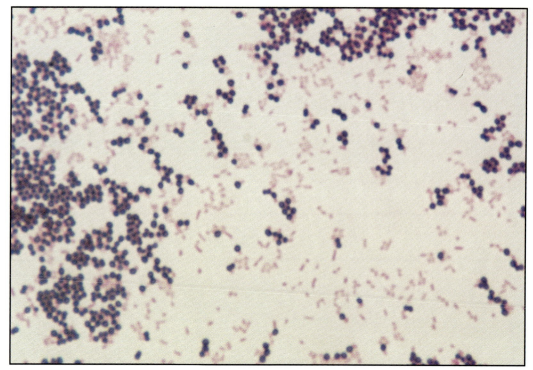

**Figure 4-2.** Gram stain of *Staphylococcus epidermidis* (+) and *Citrobacter diversus* (−) (X2640).

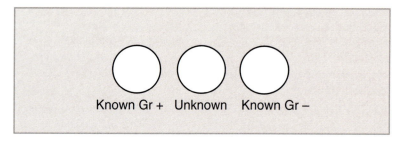

Known Gr +    Unknown    Known Gr −

**Figure 4-3.** Until consistent results are obtained, run Gram stains with known Gram-positive and Gram-negative organisms as controls.

# ACID-FAST STAIN

**Purpose**    The acid-fast stain is a differential stain used to detect cells capable of retaining a primary stain when treated with an acid alcohol.

**Medical Applications**    The acid-fast stain is an important differential stain for identifying bacteria in the genus *Mycobacterium*. *M. leprae* and *M. tuberculosis* cause leprosy and tuberculosis, respectively. Members of the actinomycete genus *Nocardia* are also acid-fast. *N. brasiliensis* and *N. asteroides* are opportunistic pathogens that cause nocardiosis, a lung disease.

**Principle**    Acid-fast cells contain a large amount of the waxy material *mycolic acid* in their walls. The wax makes it difficult for cells to adhere to a glass slide, so the smear is prepared using a drop of serum rather than water.
    Carbolfuchsin is the primary stain (see Fig. 4-4). It is a phenolic compound and is lipid soluble, so it penetrates the waxy wall. Typical aqueous stains are repelled by the waxy wall. Staining by carbolfuchsin is further enhanced by steam heating the preparation to drive the stain into the cell. Acid alcohol is used to decolorize nonacid-fast cells; acid-fast cells resist this decolorization. A counterstain of methylene blue is then applied. Acid-fast cells are reddish-purple; nonacid-fast cells are blue (see Fig. 4-5).

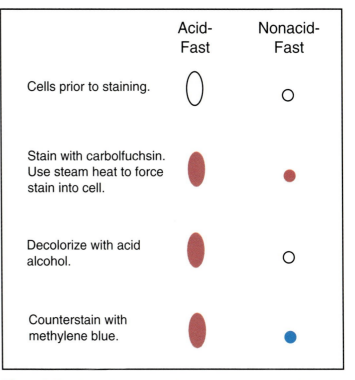

| | Acid-Fast | Nonacid-Fast |
|---|---|---|
| Cells prior to staining. | | |
| Stain with carbolfuchsin. Use steam heat to force stain into cell. | | |
| Decolorize with acid alcohol. | | |
| Counterstain with methylene blue. | | |

**Figure 4-4.** The acid-fast stain. Acid-fast cells are reddish-purple, nonacid-fast are blue.

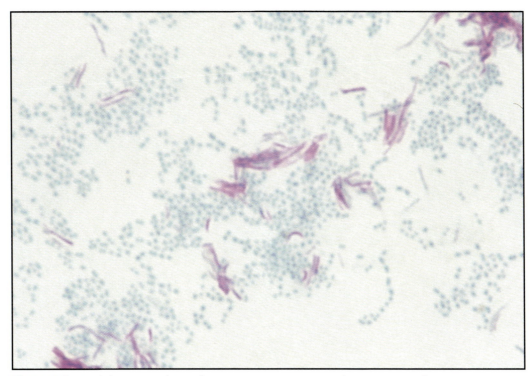

**Figure 4-5.** Acid-fast stain of *Mycobacterium smegmatis* (+) and *Enterococcus faecium* (–) (X2640).

# CAPSULE STAIN

**Purpose**   The capsule stain is a differential stain used to detect cells capable of producing an extracellular capsule.

**Medical Applications**   Capsule production increases virulence in some microbes (such as *Bacillus anthracis* and *Streptococcus pneumoniae*) by making them less vulnerable to phagocytosis. *Streptococcus mutans* produces an insoluble capsule that enables it to adhere to tooth enamel. Other bacteria become trapped in the capsule and form the plaque brushing and flossing removes. Acid production from these bacteria erodes the enamel and causes dental caries (cavities).

**Principle**   Capsules are composed of mucoid polysaccharides or polypeptides. Prior to staining, cells are emulsified in a drop of serum to promote adherence to the glass slide.

Their chemical structure makes capsules difficult to stain. Typically, the staining procedure involves an acidic stain such as Congo Red which stains the background, and a basic stain which colorizes the cell proper. The capsule remains unstained and appears as a white halo between the cells and the colored background (see Fig. 4-6).

Since the capsule stain begins as a negative stain, cells are spread in a film of the acidic stain and are not heat-fixed. Heat-fixing causes shrinkage of the cells, leaving an artifactual white halo around them that might be interpreted as a capsule.

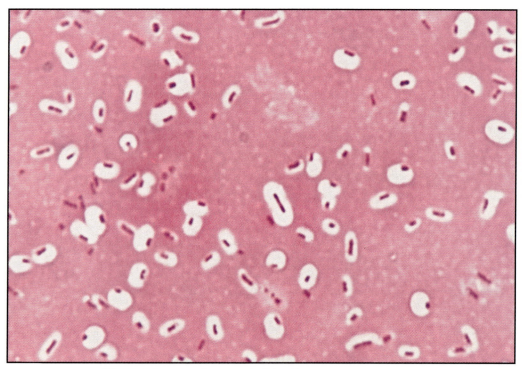

**Figure 4-6.** Capsule stain of *Klebsiella pneumoniae* (X2640).

# SPORE STAIN

**Purpose**  The spore stain is a differential stain used to detect the presence and location of spores in bacterial cells.

**Medical Applications**  Only a few genera produce spores. Among them are the genera *Bacillus* and *Clostridium*. Most members of *Bacillus* are soil, freshwater, or marine *saprophytes*, but a few are pathogens, such as *B. anthracis*, the causative agent of anthrax. Most members of *Clostridium* are soil or aquatic saprophytes, or inhabitants of human intestines, but three pathogens are fairly well known: *C. tetani* causes tetanus, *C. botulinum* causes botulism, and *C. perfringens* causes gas gangrene.

**Principle**  A spore is a dormant form of the bacterium that allows it to survive lean environmental conditions. Spores have a tough outer covering made of the protein *keratin* and are resistant to heat and chemicals. The keratin also resists staining, so extreme measures must be taken to stain the spore. In the Schaeffer-Fulton method (see Fig. 4-7), a primary stain of malachite green is forced into the spore by steaming the bacterial emulsion. Malachite green is water soluble and has a low affinity for cellular material, so vegetative cells may be decolorized with water. Vegetative cells are then counterstained with safranin.

Spores may be located in the middle of the cell (central), at the end of the cell (terminal), or between the end and middle of the cell (subterminal). These are shown in Figures 4-8 through 4-10. Spore shape may also be of diagnostic use. Spores may be spherical or elliptical (oval).

Members of the genus *Corynebacterium* may exhibit club-shaped swellings (see Fig. 4-11) that might be confused with spores. A spore stain will distinguish between true spores and these structures.

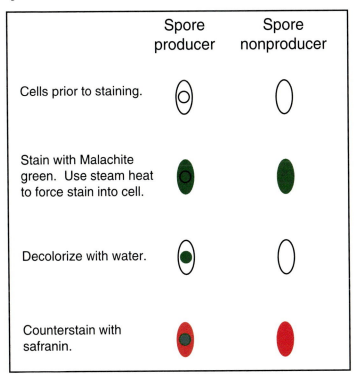

**Figure 4-7.** The Schaeffer-Fulton spore stain. Spores are green, vegetative cells red.

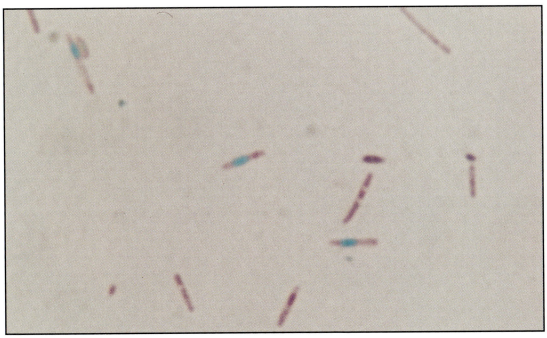

**Figure 4-8.** Central elliptical spores of *Bacillus subtilis* stained by the Schaeffer-Fulton technique (X3432).

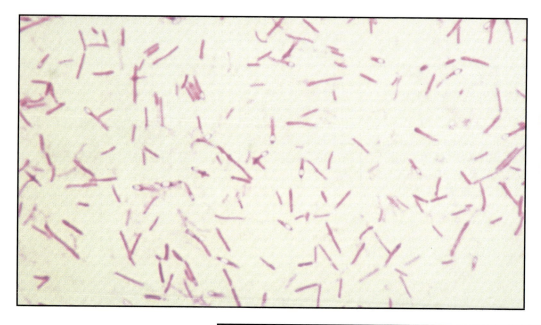

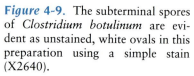

**Figure 4-9.** The subterminal spores of *Clostridium botulinum* are evident as unstained, white ovals in this preparation using a simple stain (X2640).

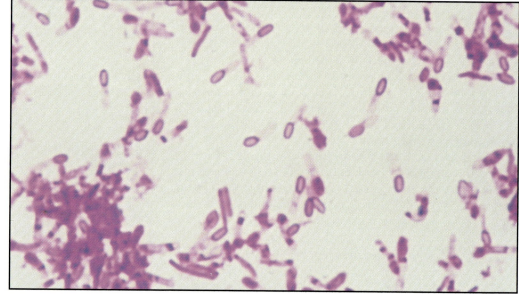

**Figure 4-10.** Spherical terminal spores of *Clostridium tetani* (X2640).

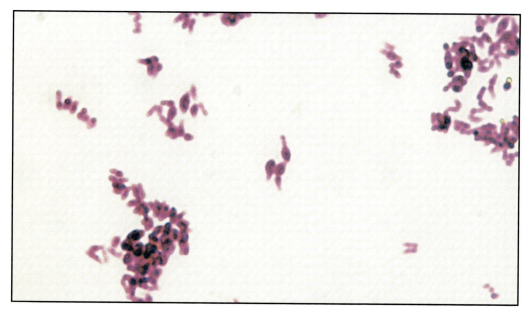

**Figure 4-11.** Club-shaped swellings of *Corynebacterium diphtheriae* should not be confused with spores (X2640).

# FLAGELLA STAIN

**Purpose**    The flagella stain allows the direct observation of flagella.

**Medical Applications**    Presence and arrangement of flagella may be useful in identifying bacterial species. Important pathogens, such as *Bordetella pertussis* (whooping cough), *Vibrio cholerae* (cholera), *Listeria monocytogenes* (meningoencephalitis), *Pseudomonas aeruginosa* (wound infections), *Salmonella typhi* (enteric fever), and *Proteus vulgaris* (urinary tract infections, bacteremia and pneumonia) are motile. Nonmotile pathogens include *Shigella dysenteriae* (bacillary dysentery), *Yersina pestis* (plague), *Neisseria gonorrhoeae* (gonorrhea) and *N. meningitidis* (meningococcal meningitis), and *Staphylococcus aureus* (abscesses, toxic shock syndrome, and others).

**Principle**    Bacterial flagella typically are too thin to be observed with the light microscope and ordinary stains. Various special flagella stains have been developed, but all require great skill and often are not performed in beginning microbiology classes. All use a mordant to assist in encrusting the flagella to a visible thickness. A variety of techniques are illustrated in the accompanying figures.

The number and arrangement of flagella may be observed with a flagella stain. A single flagellum is said to be *polar* and the cell has a *monotrichous* arrangement (see Fig. 4-12). Other arrangements (shown in Figs. 4-13 through 4-14) include *amphitrichous*, with flagella at both ends of the cell; *lophotrichous*, with tufts of flagella at the end of the cell; and *peritrichous*, with flagella emerging from the entire cell surface.

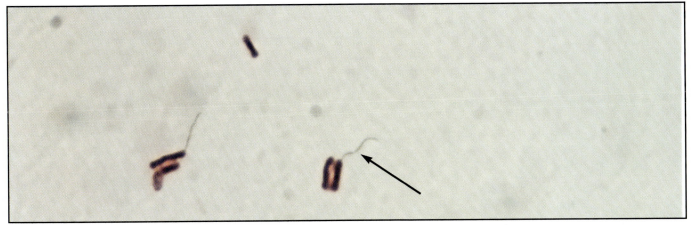

**Figure 4-12.** Polar flagella of *Pseudomonas aeruginosa* (X3432).

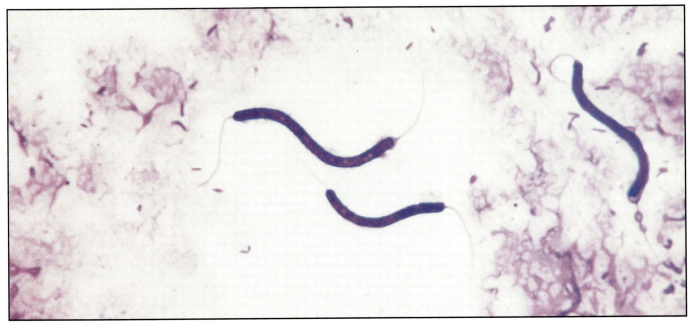

**Figure 4-13.** Amphitrichous flagella of *Spirillum volutans* (X3432).

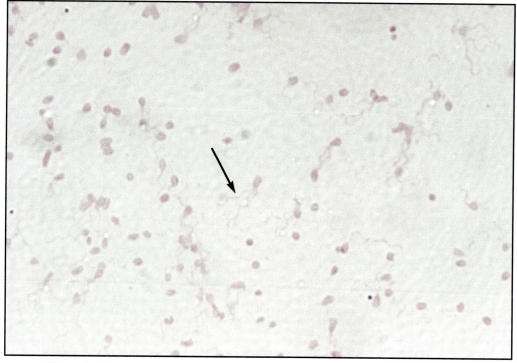

**Figure 4-14.** *"Pseudomonas reptilivora"* illustrates lophotrichous flagella (X2000).

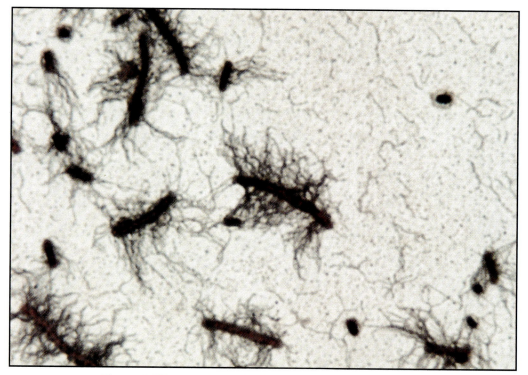

**Figure 4-15.** Peritrichous flagella of *Proteus vulgaris* (X2640).

# HANGING DROP TECHNIQUE

**Purpose**   Most bacterial microscopic preparations kill the organisms. The hanging drop technique allows observation of living cells to determine motility and cell size, arrangement and shape.

**Medical Applications**   Natural cell size, arrangement and motility may be important characteristics in the identification of a pathogen.

**Principle**   A thin ring of petroleum jelly is applied to the four edges on one side of a cover glass. A drop of water is then placed in the center of the cover glass, and living microbes are transferred into it. A depression microscope slide is carefully placed over the cover glass in such a way that the drop is received into the depression and is undisturbed.

The petroleum jelly causes the cover glass to stick to the slide, so the preparation may be picked up, inverted so the cover glass is on top, and placed under the microscope for examination. Since no stain is used and most cells are transparent, viewing is best done with as little illumination as possible. The petroleum jelly forms an air-tight seal that prevents drying of the drop allowing a long period for observation of cell size, shape, binary fission and motility.

If the technique is done to determine motility, the observer must be careful to distinguish between true motility and *Brownian motion* due to collisions with water molecules. In the latter, cells will appear to vibrate in place. With true motility, cells will exhibit independent movement over greater distances.

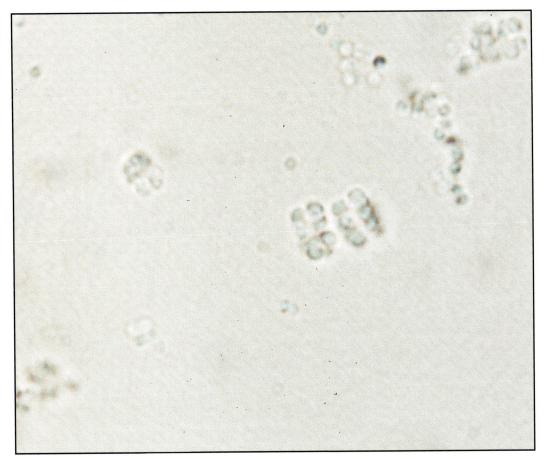

**Figure 4-16.** An unstained hanging drop preparation of *Micrococcus luteus,* a nonmotile Gram-positive coccus (X3432).

**Figure 4-17.** *Bacillus cereus* stained to show the nucleoplasm (X2640).

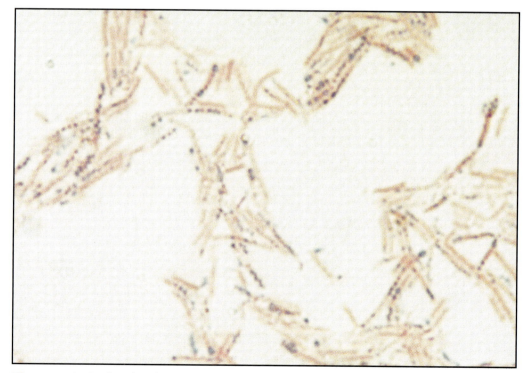

**Figure 4-18.** Dark poly-β-hydroxybutyrate (PHB) granules serve as a carbon and energy reserve (X2640 Sudan black B stain).

# Differential Tests

## BILE ESCULIN AGAR

**Purpose**   This test is used to identify the ability to hydrolyze esculin in the presence of bile.

**Medical Applications**   The bile esculin test is used to identify Group D streptococci (*Enterococcus*) which are all positive. Group D streptococci include pathogens such as *Enterococcus faecalis* and *E. faecium* (opportunistic urinary tract infections and endocarditis).

**Principle**   Bile esculin agar is both a selective and a differential medium. It is used primarily for differentiation of enterococci, so it contains not only bile and esculin, but also the Gram-negative inhibitor sodium azide. The color indicator is ferric citrate.

Esculin is a *glycoside* (a sugar molecule bonded by an acetyl linkage to an alcohol) composed of glucose and esculetin. These linkages (also called *glycosidic linkages*) are easily hydrolyzed under acidic conditions (see Fig 5-1). Many species of bacteria can hydrolyze esculin but relatively few can do so in the presence of bile.

Organisms that split esculin molecules and use the liberated glucose for their energy are also producing esculetin. Esculetin reacts with ferric citrate to form a phenolic iron complex which turns the agar slant dark brown to black (see Fig. 5-2). An agar slant that is more than half darkened after no more than 72 hours incubation is considered bile esculin positive. If less than half the slant has darkened, the result is negative (see Fig. 5-3).

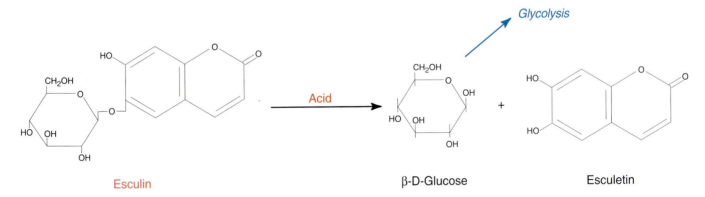

Esculin          β-D-Glucose          Esculetin

**Figure 5-1.** Acid hydrolysis of esculin with the production of esculetin.

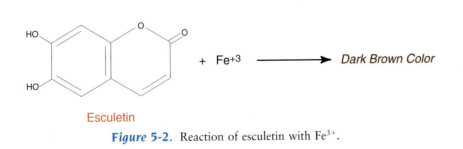

Esculetin

**Figure 5-2.** Reaction of esculetin with $Fe^{3+}$.

37

**Figure 5-3.** The bile esculin test. From left to right: *Enterococcus faecium* (+), *Serratia marcescens* (+), *Citrobacter amalonaticus* (−) and an uninoculated control. A positive result is indicated if more than half the tube is black before 72 hours.

# BLOOD AGAR

**Purpose**    Blood agar is used to determine the ability of an organism to hemolyze erythrocytes (RBCs).

**Medical Applications**    The ability to produce hemolysis is associated with virulence. Many pathogenic species of *Streptococcus*, *Staphylococcus* and *Clostridium* demonstrate hemolysis. *Streptococcus pyogenes*, which causes streptococcal sore throat, scarlet fever, rheumatic fever, and other diseases, is β hemolytic. *S. pneumoniae*, which causes pneumonia and pneumococcal meningitis, is α hemolytic. Most strains of *Staphylococcus aureus*, the causative agent of boils, toxic shock syndrome, food poisoning and other diseases, are β

hemolytic. *Clostridium botulinum* and *C. tetani* are β hemolytic and cause botulism and tetanus, respectively.

**Principle**    Exotoxins that cause destruction of RBCs (hemolysis) are called *hemolysins*. There are three categories of hemolysis. β hemolysis is the complete destruction of RBCs and hemoglobin and results in a clearing around the growth on a blood agar plate. Partial destruction of RBCs and hemoglobin (α hemolysis) produces a greenish discoloration of the blood agar plate. In γ hemolysis, there is no destruction of RBCs and hemoglobin, so there is no change in the medium.

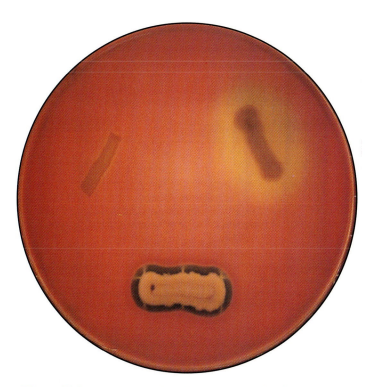

**Figure 5-4.** Blood agar plate demonstrating the three categories of hemolysis. Clockwise, from the upper left: γ hemolysis, α hemolysis, and β hemolysis.

# CATALASE TEST

**Purpose**    This test is used to identify organisms that produce the enzyme catalase.

**Medical Applications**    The catalase test may be used to distinguish between the genera *Streptococcus* (catalase −) and *Staphylococcus* (catalase +). *Streptococcus pyogenes* (streptococcal sore throat, scarlet fever, rheumatic fever, glomerulonephritis, erysipelas, and endocarditis), *S. pneumoniae* (pneumococcal pneumonia and meningitis), and *Staphylococcus aureus* (skin infections, with possible spreading to produce pneumonia, meningitis, endocarditis, and others) are the major pathogens of these two genera.

The catalase test may also distinguish between two Gram-positive, spore forming genera: catalase positive *Bacillus* and catalase negative *Clostridium*. *Bacillus anthracis* produces anthrax. *Clostridium tetani*, *C. botulinum*, and *C. perfringens*, produce tetanus, botulism, and gas gangrene, respectively.

**Principle**    Most aerobic and facultatively anaerobic bacteria that have an electron transport chain produce hydrogen peroxide via the nonenzymatic transfer of electrons from reduced flavoprotein to oxygen, as shown in Figure 5-5. Hydrogen peroxide may also be produced in aerobes and facultative anaerobes via the action of superoxide dismutase, as shown in Figure 5-5. Hydrogen peroxide is a highly reactive molecule that damages cell components. Organisms that produce the enzyme catalase are able to break hydrogen peroxide down into water and oxygen gas, as shown in Figure 5-6.

The catalase test may be performed either on a slide containing a freshly transferred pure specimen (see Fig. 5-7) or on an agar slant containing viable young colonies (see Fig. 5-8). When a drop of 3% hydrogen peroxide is placed on the growth, oxygen gas bubbles form immediately if the organism is catalase positive. If the slide test is run, observation under low power on the microscope may be useful for observing weakly positive reactions. No bubbling is considered a negative result. Hydrogen peroxide is very unstable, so a positive control should be run to verify the reagent's composition.

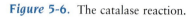

$$2H_2O_2 \xrightarrow{\text{Catalase}} 2H_2O + O_2\uparrow$$

Hydrogen Peroxide

**Figure 5-6.** The catalase reaction.

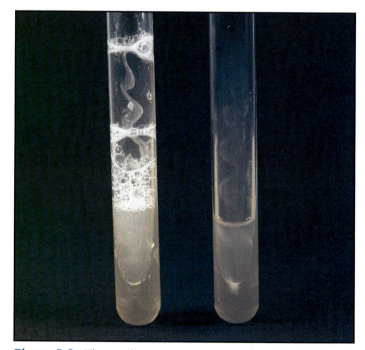

**Figure 5-7.** The catalase slide test. *Staphylococcus aureus* (+) on the left, *Enterococcus faecium* (−) on the right.

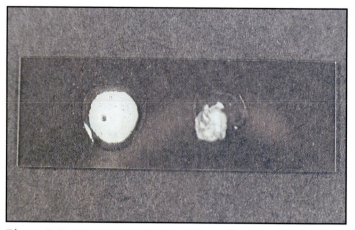

**Figure 5-8.** The catalase test on an agar slant. *Staphylococcus aureus* (+) on the left, *Enterococcus faecium* (−) on the right.

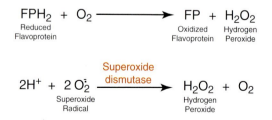

$$FPH_2 + O_2 \longrightarrow FP + H_2O_2$$

Reduced Flavoprotein         Oxidized Flavoprotein   Hydrogen Peroxide

$$2H^+ + 2O_2^- \xrightarrow{\text{Superoxide dismutase}} H_2O_2 + O_2$$

Superoxide Radical         Hydrogen Peroxide

**Figure 5-5.** Hydrogen peroxide may be formed through the transfer of hydrogens from reduced flavoprotein to oxygen or from the action of superoxide dismutase.

# CITRATE UTILIZATION AGAR

**Purpose** The citrate utilization test is used to determine the ability of an organism to use citrate as its sole carbon source. This ability depends on the enzyme *citrase*.

**Medical Applications** *Bordetella pertussis*, the causative agent of whooping cough, is citrate negative whereas all other *Bordetella* species are citrate positive. Enteric bacteria such as *Klebsiella* (*K. pneumoniae* causes pneumonia and others cause urinary tract infections) and *Enterobacter* (a nonpathogen) are usually citrate positive and may be distinguished from the opportunistic pathogen *Escherichia coli* (urinary tract infections) which is citrate negative.

**Principle** Simmon's citrate agar is a defined medium in which sodium citrate is the sole carbon source and ammonium ion is the sole nitrogen source. Bromthymol blue (BTB) is included as a pH indicator. The medium is formulated at an initial pH of 6.9 and BTB is green. At a pH more than 7.6, BTB turns a deep blue.

Citric acid typically is produced by the combination of acetyl CoA and oxaloacetic acid at the entry to the Krebs Cycle. However, some organisms are capable of using citrate as a carbon source if no fermentable carbohydrate is present. This requires citrate permease, the enzyme responsible for transporting citrate into the cell. Once inside the cell, citrate can be broken down as shown in Figure 5-9.

As $CO_2$ accumulates in the medium from citrate (+) organisms, it reacts with $Na^+$ and $H_2O$ in the Simmon's Citrate medium to produce the alkaline compound sodium carbonate ($Na_2CO_3$). This makes the medium alkaline and turns the bromthymol blue pH indicator Prussian blue (see Fig. 5-10).

Since this is an aerobic process, agar slants are used to increase the surface area exposed to air.

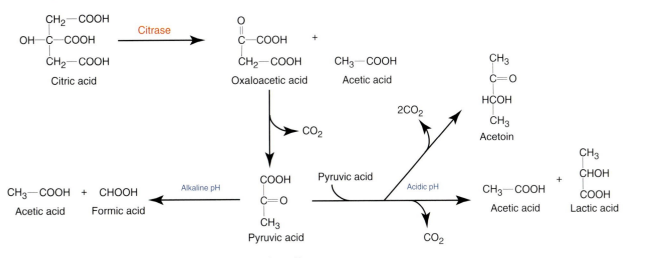

**Figure 5-9.** Metabolism in citrate (+) organisms. Once in the cell, citrate is hydrolyzed by citrase into oxaloacetic acid and acetic acid. The oxaloacetic acid is further hydrolyzed into pyruvic acid and $CO_2$. This $CO_2$ is responsible for the color change in the medium as described in the text. The fate of the pyruvic acid depends on the cellular pH as shown in the figure.

**Figure 5-10.** Simmon's citrate agar inoculated with *Citrobacter diversus* (+) on the left, *Bacillus cereus* (−) in the center, and an uninoculated control on the right.

# COAGULASE TEST

**Purpose**    The coagulase test is used to detect the ability of a suspected *Staphylococcus* to clot serum.

**Medical Applications**    This test is used to distinguish between pathogenic and nonpathogenic members of the genus *Staphylococcus*. Virtually all strains of the pathogenic *S. aureus* are coagulase positive, whereas all nonpathogenic species of the genus (*e.g. S. epidermidis*) are coagulase negative. It is thought that coagulase increases virulence by surrounding infecting organisms with a clot which protects them from host defenses, such as phagocytosis and antibodies.

**Principle**    The coagulase tube test uses citrated plasma (*i.e.* plasma treated with the anticoagulant sodium citrate to prevent normal clotting). A coagulase positive organism activates the normal clotting mechanism in some as yet unidentified way with the enzyme coagulase (staphylocoagulase). This produces a clot in the medium in as little as 30 minutes (see Fig. 5-11). Clotting may be complete or may be seen as fibrin threads. Any degree of clotting is considered a positive. Failure to clot within 24 hours is considered a negative result.

To prevent a false negative result, a positive control using *S. aureus* should be run to verify the quality of the coagulase medium. A false negative result may also occur when testing some *S. aureus* strains that produce fibrinolytic enzymes which break up the clot. To avoid a false negative, tubes should be observed for clotting every 30 minutes for the first couple of hours and the test should not be run more than 24 hours.

This test is specific to the genus *Staphylococcus*. Other genera may clot the plasma, but this is probably due to the catabolism of the anticoagulant citrate which allows the medium to coagulate rather than to coagulase.

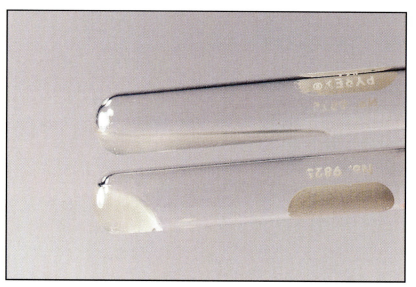

**Figure 5-11.** The coagulase tube test. Coagulase (−) *Staphylococcus epidermidis* above, coagulase (+) *S. aureus* below. This test was run for 24 hours.

# DECARBOXYLASE MEDIUM

**Purpose**   This test is used to detect the ability of an organism to decarboxylate an amino acid (typically lysine, ornithine or arginine).

**Medical Applications**   Decarboxylase tests are used to differentiate between organisms in the Enterobacteriaceae.

**Principle**   Decarboxylation is a general name given to the process of removing the carboxyl group (COOH) of an amino acid, producing an amine and carbon dioxide. Each different decarboxylase enzyme may be tested for by including its substrate in the medium. Amino acids commonly used for clinical identification are lysine, arginine and ornithine. Decarboxylation in general is shown in Figure 5-12 and the specific reactions are shown in Figures 5-13 through 5-15.

Møller's decarboxylase medium contains peptone, glucose, bromcresol purple, and a coenzyme pyridoxal phosphate in addition to the specific amino acid substrate. An overlay of mineral oil is used to limit oxygen in the medium and promote fermentation with the accumulation of acid end products since decarboxylase enzymes are inducible and are only produced in the presence of their substrate and an acidic environment.

Bromcresol purple is a pH indicator. It is purple at pH greater than 6.8 and yellow at pH less than 5.2. Glucose fermentation in the anaerobic medium initially turns it yellow due to the accumulation of acid end products. The low pH induces decarboxylase (+) organisms to produce the enzyme. Subsequent decarboxylation turns the medium purple due to the accumulation of alkaline end products (see Fig. 5-16). If the organism is only capable of glucose fermentation, the medium will remain yellow.

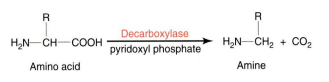

**Figure 5-12.** Decarboxylation of an amino acid.

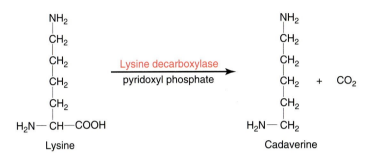

**Figure 5-13.** Decarboxylation of lysine.

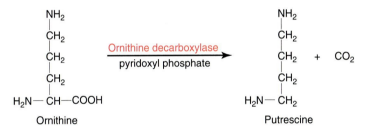

**Figure 5-14.** Decarboxylation of ornithine.

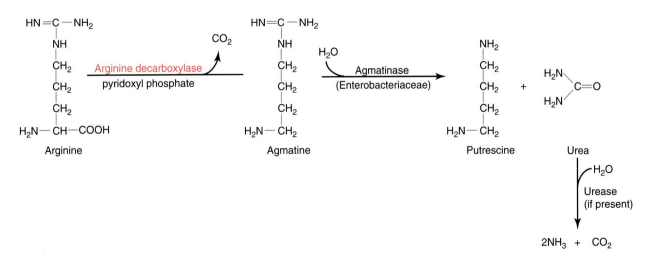

**Figure 5-15.** Decarboxylation of the amino acid arginine produces the amine agmatine. Members of the Enterobacteriaceae are capable of degrading agmatine into putrescine and urea. Those strains with urease can further break down the urea into ammonia and carbon dioxide. Thus, the end products of arginine catabolism are carbon dioxide, putrescine and urea, or carbon dioxide, putrescine and ammonia.

**Figure 5-16.** Lysine decarboxylase test. *Proteus vulgaris* (−) on the left and *Klebsiella pneumoniae* (+) on the right. An uninoculated control is in the center.

*A Photographic Atlas for the Microbiology Laboratory*

# DNASE TEST AGAR

**Purpose**   DNase test agar is used to identify bacteria capable of producing the exoenzyme DNase.

**Medical Applications**   DNase test agar is used in combination with other tests (e.g. mannitol salt agar, coagulase, blood agar, and tellurite glycine agar) to identify the presence of pathogenic *Staphylococcus aureus*. Group A streptococci (mainly *Streptococcus pyogenes* which may cause erysipelas, puerperal fever, sepsis, streptococcal sore throat and endocarditis) and *Serratia marcescens* (an opportunistic pathogen) also produce extracellular DNase.

**Principle**   An enzyme that catalyzes the depolymerization of DNA into small fragments is called a *deoxyribonuclease* or *DNase* (see Fig. 5-17). Ability to produce this enzyme can be determined by culturing and observing a suspected organism on a DNase test agar plate.

One type of DNase test agar contains an emulsion of DNA and methyl green dye in a complex that gives the agar a blue-green color. Bacterial colonies that secrete DNase will hydrolyze the DNA in the medium resulting in clearing around the growth (see Fig. 5-18).

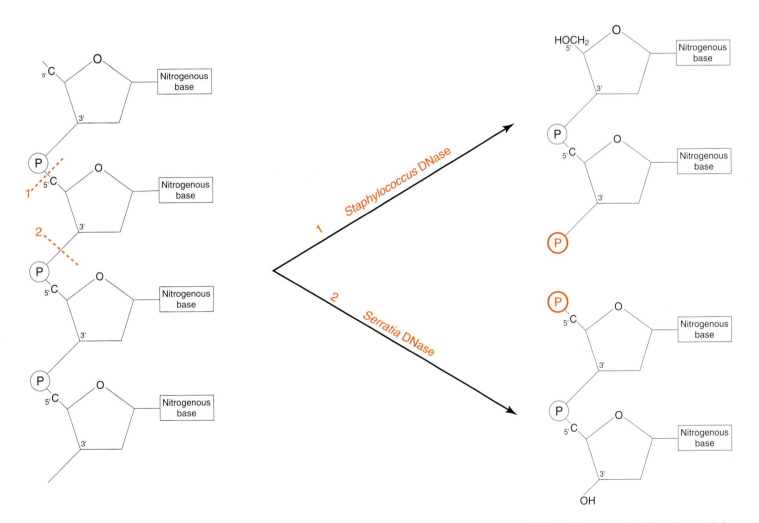

**Figure 5-17.**   Two patterns of DNA hydrolysis. DNase from *Staphylococcus* hydrolyzes DNA at the bond between the 5'-carbon and the phosphate (illustrated by line 1), thereby producing fragments with a free 3'-phosphate (shown in red on the upper fragment). Most fragments are one or two nucleotides long. A dinucleotide is shown here. *Serratia* DNase cleaves the bond between the phosphate and the 3'-carbon (illustrated by line 2) and produces fragments with free 5'-phosphates (shown in red on the lower fragment). Most fragments are two to four nucleotides in length. A dinucleotide is shown here.

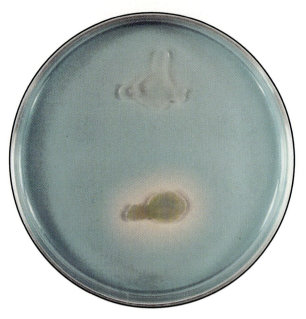

Figure 5-18. DNase agar with methyl green. *Staphylococcus aureus* (+) below and *S. epidermidis* (−) above.

*A Photographic Atlas for the Microbiology Laboratory*

# ENTEROTUBE® II

**Purpose**  The Enterotube® II is used for identification of enteric bacteria.

**Medical Applications**  Many enteric bacteria are pathogens. Among them are *Shigella dysenteriae* (bacillary dysentery), *Salmonella typhi* (typhoid fever), *Proteus mirabilis* (opportunistic urinary tract infections), *Escherichia coli* (opportunistic urinary tract infections) and *Yersinia pestis* (plague).

**Principle**  The Enterotube® II consists of 12 compartments containing various media (the tests are shown in Fig. 5-19) and allows determination of 15 different characteristics of the organism. (The principles behind each individual test are covered elsewhere in this manual.) The Enterotube® II contains a wire which is touched to a colony of the organism to be tested. The wire is then drawn through the tube inoculating the media in each chamber as it passes. After 18 to 24 hours incubation, the test results are read by comparing the inoculated tube to an uninoculated control. Two inoculated tubes and a control are shown in Figure 5-20.

Identification of the enteric bacterium requires scoring all tests but VP. The remaining tests are clustered into groups, and each positive result within a group is assigned a number—either 4, 2 or 1 (see Fig. 5-21). The numbers within each group are added together, with their sums being used to make a five digit number. Thus, the combination of test results gives a unique (usually) ID value for each organism. The ID value is then located in the Computer Coding and Identification System (CCIS) to give the identity of the unknown organism. Should more than one organism have the same ID value, a test (often VP) to differentiate between them is suggested.

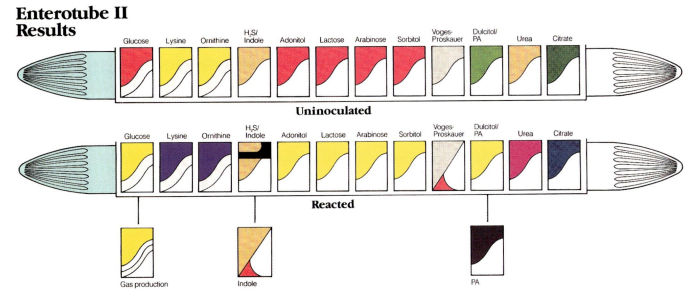

**Figure 5-19.**  Enterotube® II results. (Illustration courtesy Becton Dickinson and Company.)

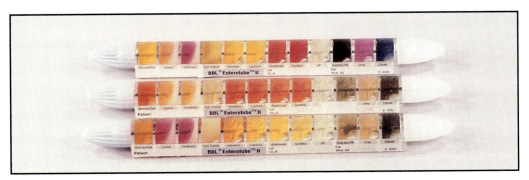

**Figure 5-20.**  Two inoculated tubes in front of and behind an uninoculated control tube.

**ENTEROTUBE® II**

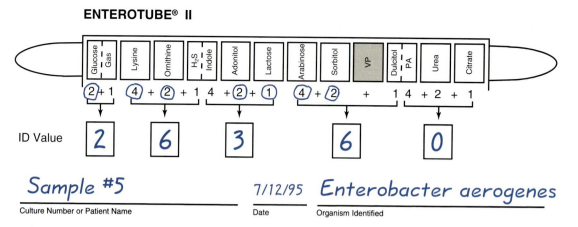

Sample #5 — Culture Number or Patient Name

7/12/95 — Date

*Enterobacter aerogenes* — Organism Identified

**Figure 5-21.** A sample score sheet for the front tube in Figure 5-20. In this case, the organism was positive for glucose fermentation, lysine decarboxylase, ornithine decarboxylase, adonitol fermentation, and lactose fermentation. The ID value obtained was 26360, which corresponds to *Enterobacter aerogenes*.

# GELATIN LIQUEFACTION TEST (NUTRIENT GELATIN)

**Purpose**   This test is used to determine the ability of a microbe to produce hydrolytic exoenzymes called *gelatinases* that digest and liquefy gelatin.

**Medical Applications**   Gelatin liquefaction may be used to distinguish between the pathogenic *Staphylococcus aureus* (+) and nonpathogenic *S. epidermidis* (slow +). It may also be useful in distinguishing *Listeria monocytogenes* (−), one causative agent of bacterial meningitis, from some species of *Corynebacterium*.

**Principle**   Many nutrient sources are too large to enter the cell. Some bacteria have the ability to produce and secrete enzymes that hydrolyze these compounds into smaller subunits that the cell can use.

Gelatin is a protein derived from collagen, a connective tissue found in vertebrates. Bacterial hydrolysis of gelatin occurs in two sequential reactions catalyzed by a family of exoenzymes referred to as *gelatinases*. These reactions are shown in Figure 5-22. Amino acids may then be used as an energy source for the cell or built back up into bacterial protein.

Nutrient gelatin tubes are stab inoculated, then incubated for up to a week. Gelatinase (+) organisms will liquefy the medium; the medium will remain solid when inoculated with gelatinase (−) organisms (see Fig. 5-23). Care must be taken to distinguish between gelatin hydrolysis and gelatin melting. An uninoculated control should be incubated with the inoculated tubes since nutrient gelatin melts at 28°C. If the control is liquid after incubation, all tubes should be refrigerated until the control is solid.

**Figure 5-22.**  Hydrolysis of gelatin by the gelatinase group of enzymes.

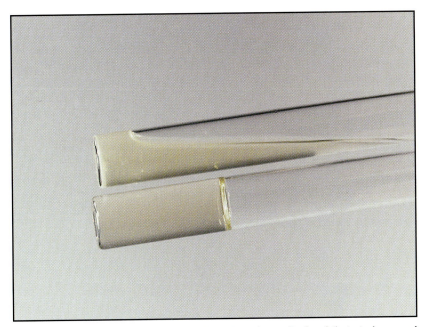

**Figure 5-23.**  Nutrient gelatin stab tubes. *Aeromonas hydrophila* (+) above and *Micrococcus roseus* (−) below.

# KLIGLER'S IRON AGAR (KIA)

**Purpose**    Kligler's iron agar (KIA) differentiates bacteria based on their ability to ferment glucose and/or lactose, and to produce hydrogen sulfide.

**Medical Applications**    KIA is used to distinguish between members of the Enterobacteriaceae, all of which ferment glucose to an acid end product. Because they are so similar morphologically, a single medium providing a variety of results is of great use in their identification. Many members of this group are pathogenic, including *Escherichia coli* (opportunistic urinary infections), *Salmonella typhi* (typhoid fever), *Shigella* (bacillary dysentery), *Yersinia pestis* (plague), and *Klebsiella* (pneumonia).

**Principle**    Kligler's iron agar test is different from other tests in many ways: the way the medium is prepared, the method of its inoculation, the way it is read, and particularly the many different interpretations that can be obtained from it.

The various reactions in KIA are due not only to the ingredients, but to the relative proportions of the sugars to each other and to the peptone. The medium is formulated with 2% polypeptone, 1% lactose and 0.1% glucose. Thiosulfate is included as an electron acceptor for sulfur reducers and ferric ammonium citrate is added as the $H_2S$ indicator. Phenol red (yellow at pH less than 6.8 and red above 6.8) is the pH indicator.

KIA medium is prepared as an agar slant, but it differs in that the butt of the agar is approximately 4 cm deep (rather than the more typical 2 cm). It is inoculated with an inoculating needle (not a loop) first by a stab in the agar butt followed by a fishtail streaking of the slant.

Many interpretations can be made from the results of a KIA test (see Fig. 5-24). The test must be read 18 to 24 hours after inoculation to avoid obtaining false acid or alkaline results. All test results include one or more of the following: glucose fermentation, lactose fermentation, gas production, no fermentation, alkaline reaction or hydrogen sulfide production. Standard symbols for reporting KIA results are as follows: A = acid, alk = alkaline, G = gas, $H_2S$ = sulfur reduction positive, and NC = no change. These symbols are written with the slant results first followed by a slash and the butt results. The various results are described below and shown in Figure 5-25.

Glucose fermentation with acid production is indicated by a yellow color throughout the medium within a few hours. By 18 to 24 hours, the slant may have turned red while the butt remains yellow. This is because the bacteria that preferentially use glucose have oxidatively exhausted it and have begun to metabolize the peptone. This produces $NH_3$ and raises the pH, giving an alkaline result. The butt remains yellow because the acid has not been neutralized by the peptone breakdown products.

If the organism ferments lactose, it will also ferment glucose. However, because of the high lactose concentration, fermentation will not be complete after 24 hours and both the slant and butt will be yellow. An early reading could give a false lactose acid positive result. After exhaustion of the lactose, the slant will turn red due to peptone metabolism. A late reading, therefore, could give a false lactose negative result.

Gas production can occur whenever fermentation is taking place. Gas produced in KIA tubes will appear as pockets or fissures in the medium or will lift the entire agar butt off the bottom of the tube.

An organism that does not ferment glucose or lactose may utilize the peptone and turn the medium red. If the organism can catabolize the peptone aerobically and anaerobically, both the slant and butt will turn red. If the organism is a strict aerobe, only the slant will turn red.

Hydrogen sulfide ($H_2S$) may be produced by the reduction of thiosulfate in the medium or by the breakdown of cysteine in the peptone. Ferric ammonium citrate reacts with the $H_2S$ to form a black precipitate, usually in the butt of the agar. If the black precipitate obscures the color of the medium, it is recorded as acid positive, since an acid environment must exist for sulfur reduction to occur.

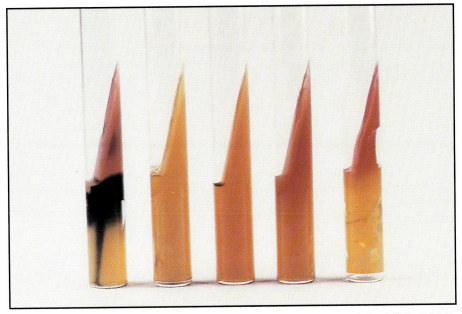

**Figure 5-24.** KIA agar slants, from left to right: *Proteus mirabilis* (alk/A,G,H₂S), *Escherichia coli* (A/A,G), uninoculated control, *Pseudomonas aeruginosa* (alk/NC), and *Morganella morganii* (alk/A,G).

| RESULTS | SYMBOL | INTERPRETATION |
|---|---|---|
| red/yellow | alk/A | glucose fermentation only<br>peptone catabolized |
| yellow/yellow | A/A | glucose and lactose fermentation |
| red/red | alk/alk | no fermentation<br>peptone catabolized |
| red/no color change | alk/NC | no fermentation<br>peptone used aerobically |
| yellow/yellow with bubbles | A/A,G | glucose and lactose fermentation<br>gas produced |
| red/yellow with bubbles | alk/A,G | glucose fermentation only<br>gas produced |
| red/yellow with bubbles and black precipitate | alk/A,G, H₂S | glucose fementation only<br>gas produced<br>H₂S produced |
| red/yellow with black precipitate | alk/A, H₂S | glucose fermentation only<br>H₂S produced |
| yellow/yellow with black precipitate | A/A, H₂S | glucose and lactose fermentation<br>H₂S produced |
| yellow/yellow with bubbles and black precipitate | A/A, G, H₂S | glucose and lactose fermentation<br>gas produced<br>H₂S produced |
| no change/no change | NC/NC | no fermentation (organism is growing very slowly or not at all) |

**Figure 5-25.** Results, symbols and interpretations of KIA agar.

# LITMUS MILK MEDIUM

**Purpose**    Bacteria may be differentiated based on the variety of reactions in litmus milk.

**Medical Applications**    Stormy fermentation of litmus milk is useful in identification of *Clostridium perfringens*, the causative agent of gas gangrene.

**Principle**    Litmus milk is an undefined medium consisting of skim milk and litmus. Skim milk provides nutrients for growth, with lactose as the carbohydrate source and casein as the primary protein source. Litmus is a pH indicator that is pink at an acid pH (4.5) and blue at an alkaline pH (8.3). Between these extremes, it is purple. (The medium is formulated to an initial pH of 6.8, thus accounting for its purple color.)

Four basic categories of reaction may occur, either separately, or in combination. These are: lactose fermentation, reduction of litmus, casein hydrolysis and casein coagulation. Figures 5-26 and 5-27 illustrate some reactions in litmus milk. Figure 5-31 summarizes the reactions in litmus milk.

Lactose is a disaccharide which yields the monosaccharides glucose and galactose when hydrolyzed by the enzyme β-galactosidase (see Fig. 5-28). Glucose may then be fermented to acid end products, lowering the pH and turning the litmus a pink color. This is an *acid reaction*. Accumulating acid may also cause precipitation of casein and form an *acid clot* (see Fig. 5-29). If *gas* is produced, fissures may be visible in the clot. Extreme gas production may break the clot and produce *stormy fermentation*.

Some bacteria have proteolytic enzymes such as rennin, pepsin or chymotrypsin that digest casein and coagulate (curdle) the milk. The action of rennin (also known as *rennet*) is shown in Fig. 5-30. The insoluble rennet curd is soft and retracts from the sides of the tube (unlike an acid clot) leaving behind a grayish fluid called whey. This reaction is a *soft clot* or *soft curd*. *Proteolysis* of the soft curd occurs if bacteria have the appropriate proteolytic enzymes to break the protein down into its component amino acids. Proteolysis is evidenced by a brownish fluid.

Some bacteria grow in litmus milk but do not ferment the lactose. Rather, they partially digest the casein. This releases $NH_3$ which raises the pH and the litmus turns blue. This is an *alkaline* reaction.

Litmus may act as the electron acceptor during lactose fermentation. Reduction of litmus is evidenced by its conversion to a white color.

**Figure 5-26.** Reactions in litmus milk. From left to right: *Pseudomonas aeruginosa* (P), *Streptococcus lactis* (ACR), *Klebsiella pneumoniae* (AGCR-note small fissure in clot), *Aeromonas hydrophila* (CR), uninoculated control, *Staphylococcus aureus* (A), and *Alcaligenes faecalis* (alk).

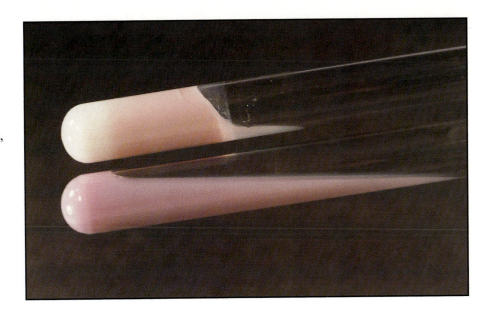

**Figure 5-27.** An acid clot in the top tube, an uninoculated control below.

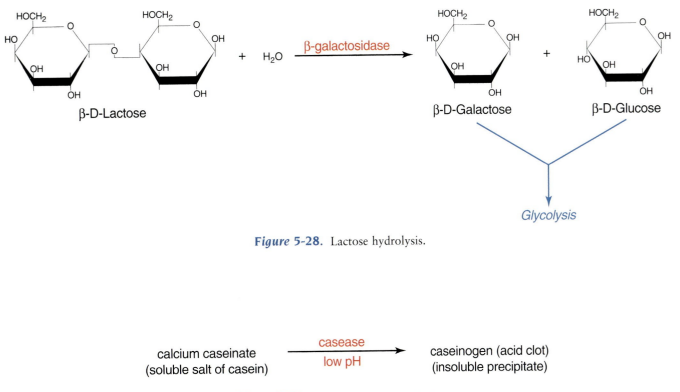

**Figure 5-28.** Lactose hydrolysis.

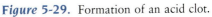

calcium caseinate          caseinogen (acid clot)
(soluble salt of casein)   (insoluble precipitate)

caseease / low pH

**Figure 5-29.** Formation of an acid clot.

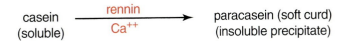

casein          paracasein (soft curd)
(soluble)       (insoluble precipitate)

rennin / Ca⁺⁺

**Figure 5-30.** Rennin activity in the formation of a curd (soft clot).

| RESULTS | SYMBOL | INTERPRETATION |
| --- | --- | --- |
| purple (same as control) | NC | no reaction |
| pink throughout | A | lactose fermentation with acid end product |
| pink at surface, white below | AR | lactose fermentation with acid end product<br>reduction of litmus |
| pink at surface, white below and solid | ACR | lactose fermentation with acid end product<br>acid clot<br>reduction of litmus |
| pink at surface, white below and solid with fissures | AGCR | lactose fermentation with acid *and* gas end products<br>acid clot<br>reduction of litmus |
| pink at surface, white below, clot broken apart | AGCS | lactose fermentation with acid *and* gas end products<br>(stormy fermentation)<br>acid clot<br>reduction of litmus |
| blue | alk | alkaline reaction |
| semisolid | C | rennet curd |
| solid material at bottom, brownish fluid above | P | proteolysis (peptonization) of casein |
| white below, purple at top | R | reduction of litmus |

*Figure 5-31.* A summary of common litmus milk reactions.

# METHYL RED AND VOGES PROSKAUER (MRVP) BROTH

**Purpose**   The methyl red test is used to identify bacteria that ferment glucose to stable acid end products using a mixed-acid fermentation The Voges-Proskauer test is used to identify bacteria capable of a 2,3-butanediol fermentation as a continuation of mixed-acid fermentation.

**Medical Applications**   *Yersinia pestis* (plague), *Listeria monocytogenes* (meningitis), *Salmonella typhi* (typhoid fever), *Shigella dysenteriae* (bacillary dysentery) and *Escherichia coli* (opportunistic urinary tract infections) are MR (+), whereas *Enterobacter aerogenes* (urinary tract infections) is MR(−). *Staphylococcus aureus* (bacteremia, pneumonia and endocarditis) and *Klebsiella pneumoniae* (pneumonia) are VP (+).

**Principle**   MRVP broth includes peptone, glucose and a phosphate buffer. Some bacteria (especially enterics) perform a *mixed-acid fermentation* (see Fig. 5-32) and are able to ferment the glucose and produce enough stable acid end products to overcome the buffering system and lower the pH. These organisms are methyl red positive. However, many microbes produce acids within 18-24 hours and then catabolize them further to more neutral compounds. To avoid reading a false positive, the methyl red test must be run for 2 to 5 days to assure the presence of *stable* acids.

After an appropriate incubation time, a small aliquot of sample is removed and methyl red indicator is added. Methyl red is red at a pH less than 4.4 and yellow at a pH greater than 6.0. Between these two pH values, methyl red is various shades of orange. A red color indicates a positive methyl red result and yellow is negative (see Fig. 5-33). An orange color indicates further incubation is required.

The enteric bacteria perform mixed-acid fermentation. However, not all enterics produce stable acid end products from it. Some further metabolize the acids to less acidic end products, such as 2,3-butanediol (see Fig. 5-34). The Voges-Proskauer test is used to detect this ability. Reagents used in this test don't actually identify 2,3-butanediol, but rather identify its precursor acetoin (acetylmethylcarbinol). Since these two substances are always found together, the presence of one is a reliable indicator for the presence of the other.

After appropriate incubation, Barritt's Reagents A (α-naphthol) and B (KOH) are added to the sample. The tube is gently shaken to aerate the medium and oxidize any acetoin to diacetyl (see Fig. 5-35). Diacetyl further reacts with components of the peptone (nitrogenous compounds called *guanidine nuclei*) to form a red color. Red is therefore a positive result for the VP test (see Fig. 5-36). No color change or a copper color (due to the reaction of KOH and α-naphthol) are negative results.

One might expect MR and VP results to be complementary. That is, MR (+) organisms should be VP (−) and VP (+) organisms should be MR (−). However, this is not always the case as *Hafnia alvei* and *Proteus mirabilis* are both typically MR (+) *and* VP (+).

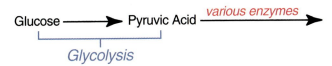

Lactic acid ($CH_3CHOH \cdot COOH$)
Carbon dioxide ($CO_2$)
Hydrogen gas ($H_2$)
Ethanol ($CH_3CH_2OH$)
Acetic acid ($CH_3COOH$)
Succinic acid ($HOOC \cdot CH_2CH_2COOH$)
Formic acid ($HCOOH$)

**Figure 5-32.** The mixed-acid fermentation of *Escherichia coli*, a representative methyl red positive enteric bacterium. Products are listed in order of abundance. Most of the formic acid is converted to $H_2$ and $CO_2$ gases.

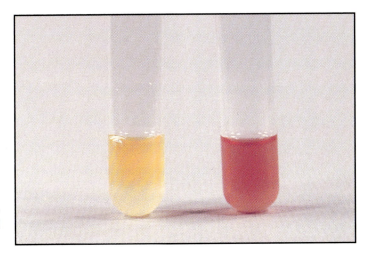

**Figure 5-33.** The methyl red test. *Enterobacter aerogenes* (MR −) on the left and *Escherichia coli* (MR +) on the right.

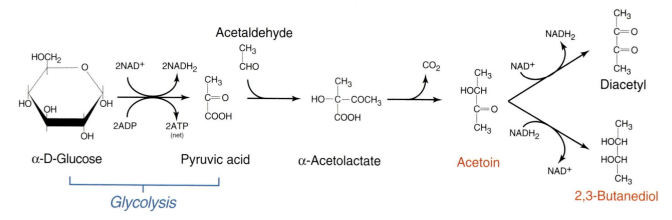

Figure 5-34. A 2,3-butanediol fermentation.

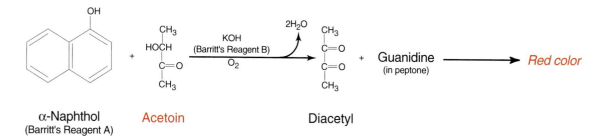

Figure 5-35. Chemistry of the Voges-Proskauer test.

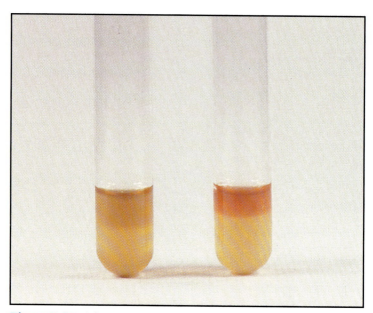

Figure 5-36. The Voges-Proskauer test. *Escherichia coli* (VP −) on the left and *Klebsiella pneumoniae* (VP+) on the right. The copper color at the top of the VP (−) tube is due to the reaction of KOH and α-naphthol and should not be confused with a positive result.

# MILK AGAR

**Purpose**    Milk agar is used to detect the presence of the proteolytic exoenzyme *casease*.

**Medical Applications**    Casease production is not a diagnostic feature of any pathogenic group. However, a number of pathogens are casease positive, including *Kingella indologenes* (eye infections), *Flavobacterium meningosepticum* (neonatal meningitis), *Bacteroides bivius* and *B. disiens* (opportunistic urogenital or abdominal infections) and *Chromobacterium violaceum* (opportunistic infections of the gut, sepsis and abscesses).

**Principle**    Many extracellular carbon compounds are too large to enter the cell. Some cells are able to secrete *exoenzymes* that hydrolyze these compounds into smaller, soluble molecules that are able to enter the cell. Enzymes that hydrolyze protein are called *proteases* (see Fig. 5-37).

Milk agar is an undefined medium containing peptone, beef extract and *casein* (milk protein). Casein is a colloid in the milk and is responsible for giving milk its white color. Organisms that produce casease digest casein and will show a zone of clearing around their growth. Thus, clearing of the medium is interpreted as a positive result and no clearing is a negative (see Fig. 5-38).

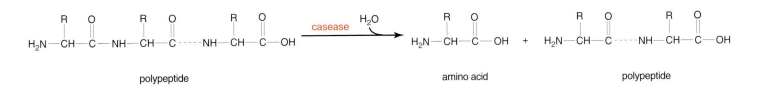

polypeptide                                                                         amino acid                          polypeptide

**Figure 5-37.**  Protein hydrolysis occurs by breaking peptide bonds (one is shown by the red line) between adjacent amino acids to produce short peptides or individual amino acids.

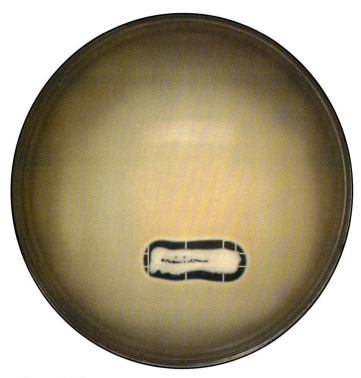

**Figure 5-38.**  Milk agar with *Bacillus coagulans* (−) above and *B. megaterium* (+) below.

# MOTILITY AGAR

**Purpose**    This test is used to detect motility in bacteria.

**Medical Applications**    Motility may be useful in identifying pathogens. Among the motile pathogens are: *Pseudomonas aeruginosa* (opportunistic wound infections, meningitis and urinary infections), *Vibrio cholerae* (cholera), *Salmonella typhi* (typhoid fever), *Proteus spp.* (urinary tract infections) and *Aeromonas hydrophila* (diarrhea).

**Principle**    A motility agar tube is inoculated by stabbing with a straight transfer needle. Since the medium is formulated with a lower agar concentration than used in nutrient agar, motile bacteria will exhibit growth outwards from the stab line.

A tetrazolium salt (TTC) may be included in the medium to make interpretation easier. TTC is used by the bacteria as an electron acceptor. In its oxidized form, TTC is colorless and soluble; when reduced it is red and insoluble (see Fig. 5-39). A positive result for motility is indicated when the red TTC is found throughout the medium. A negative result shows red only along the stab line. Figures 5-40 and 5-41 illustrate motility agar with and without TTC.

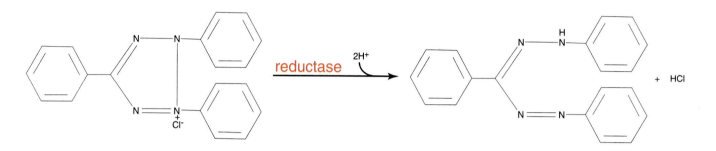

2,3,5-Triphenyltetrazolium chloride (TTC)                Formazan (red color)

**Figure 5-39.** Reduction of the colorless and soluble 2,3,5-Triphenyltetrazolium chloride by metabolizing bacteria results in its conversion to a red and insoluble formazan. The location of the growing bacteria can be easily determined by the location of the red color in the medium.

**Figure 5-40.** Motility agar tubes containing TTC inoculated with *Aeromonas hydrophila* (motile) on the left, *Micrococcus luteus* (nonmotile) in the center, and an uninoculated control on the right.

**Figure 5-41.** Motility agar tubes without TTC inoculated with *Aeromonas hydrophila* (motile) on the left, an uninoculated control in the center, and *Micrococcus luteus* (nonmotile) on the right. Compare with Figure 5-40.

# NITRATE REDUCTION BROTH

**Purpose**    This test is used to detect the ability of an organism to reduce nitrate ($NO_3$) to nitrite ($NO_2$) or some other nitrogenous compound, such as molecular nitrogen ($N_2$).

**Medical Applications**    Most enteric bacteria are nitrate reducers, including pathogenic species such as *Escherichia coli* (opportunistic urinary tract infections), *Klebsiella pneumoniae* (pneumonia), *Morganella morganii* and *Proteus vulgaris* (nosocomial infections), *Proteus mirabilis* (opportunistic urinary tract infections), *Salmonella typhi* (enteric or typhoid fever), *Shigella dysenteriae* (bacillary dysentery) and most strains of *Yersina pestis* (plague). Among nonenteric nitrogen reducing pathogens are *Staphylococcus aureus* (staphylococcal food poisoning, staphylococcal bacteremia, various abscesses) and *Bacillus anthracis* (anthrax).

**Principle**    Nitrate ($NO_3$) may be reduced to several different compounds (shown in Fig. 5-42) via two metabolic processes. These are *anaerobic respiration* and *denitrification*. During anaerobic respiration, the organism uses nitrate as the final electron acceptor of the cytochrome system, producing nitrite, ammonia, molecular nitrogen, nitric oxide, or some other reduced nitrogenous compound, depending on the species. Denitrification reduces nitrate to molecular nitrogen This is an essential component of the nitrogen cycle.

Nitrate reduction broth is an undefined medium. In addition to essential nutrients and a nitrate source ($KNO_3$), it has a small amount of agar added to slow oxygen diffusion and encourage anaerobic growth.

After incubation, reagents containing sulfanilic acid and α-naphthylamine are added to detect the presence of nitrite (phase 1 reaction). If nitrite is present, it reacts to form nitrous acid ($HNO_2$) which reacts with the sulfanilic acid and α-naphthylamine to produce a red, water soluble compound (see Fig. 5-43). A red color, therefore, indicates the presence of nitrite and is interpreted as a positive result for nitrate reduction (see Fig. 5-44).

If no color change occurs, it could be the organism doesn't reduce nitrate, or it could be that the organism reduces nitrate to some compound other than nitrite. To discriminate beween these two possibilities, zinc dust is added to the medium (phase 2 reaction). If nitrate remains in the medium, zinc will reduce it to nitrite, and a pink color will be observed. *A pink color upon addition of zinc is interpreted as a negative for nitrate reduction.* No color change means nitrate has been reduced and is interpreted as a positive (see Fig. 5-45).

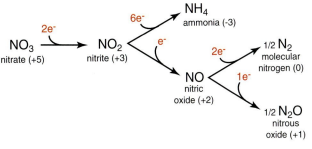

**Figure 5-42.** Possible end products of nitrate reduction. The oxidation state of nitrogen in each compound is shown in parentheses.

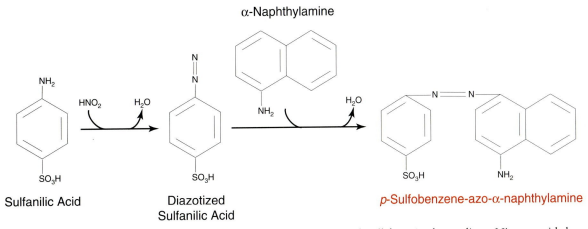

**Figure 5-43.** Phase 1 indicator reaction. If nitrate is reduced to nitrite, nitrous acid will form in the medium. Nitrous acid then reacts with sulfanilic acid to form diazotized sulfanilic acid, which reacts with the α-naphthylamine to form *p*-sulfobenzene-azo-α-naphthylamine, which is red. Thus, a red color indicates the presence of nitrite and is considered a positive result for nitrate reduction.

**Figure 5-44.** Nitrate reduction broth tubes after addition of sulfanilic acid and α-naphthylamine. From left to right: *Klebsiella pneumoniae* (+1), an uninoculated control, *Enterococcus faecium* (−), and *Pseudomonas aeruginosa* (−). A negative result at this point must be checked for the presence of nitrate in the medium; a positive indicates reduction of nitrate to nitrite.

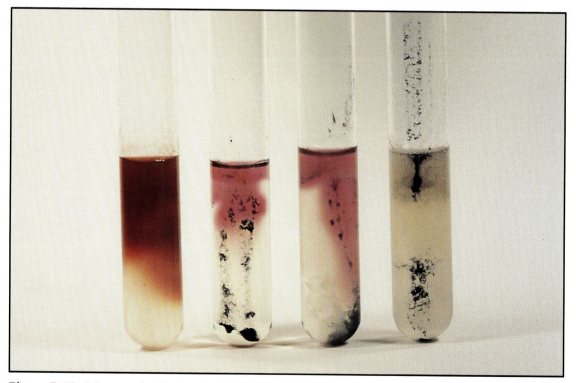

**Figure 5-45.** Nitrate reduction broth tubes after addition of zinc. A red color indicates the presence of nitrate, which means it wasn't reduced. No color change is considered a positive result for nitrate reduction. From left to right: *Klebsiella pneumoniae* (no zinc added since it was positive in phase 1), an uninoculated control, *Enterococcus faecium* (−), and *Pseudomonas aeruginosa* (+2).

# o-NITROPHENYL-β-D-GALACTOPYRANOSIDE (ONPG) TEST

**Purpose**   The ONPG test is used to determine the presence of the enzyme β-galactosidase which hydrolyzes β-lactose to β-glucose and β-galactose.

**Medical Applications**   The ability to hydrolyze β-lactose is a characteristic of some pathogenic bacteria, such as *Escherichia coli* (opportunistic urinary infections), *Yersinia pestis* (plague), and *Klebsiella pneumoniae* (pneumonia). Some other important pathogens characteristically give a negative result for this test: *Neisseria gonorrhoeae* (gonorrhea), *Neisseria meningitidis* (meningitis), and *Pseudomonas aeruginosa* (opportunistic infections of burn victims).

**Principle**   For bacteria to ferment lactose, two enzymes must be present: 1) β-galactoside permease, a membrane-bound transport protein, and 2) β-galactosidase, an intracellular enzyme that splits the disaccharide into glucose and galactose (see Fig. 5-46).

Bacteria possessing both enzymes are active β-lactose fermenters. Bacteria possessing neither enzyme never ferment β-lactose. Bacteria that possess β-galactosidase but no β-galactoside permease may mutate and, over a period of days or weeks, begin to produce the permease. This third group, called *late lactose fermenters,* can be differentiated from the nonfermenters using a compound called o-nitrophenyl-β-D-galactopyranoside (ONPG). This compound can enter the cell without the assistance of β-galactoside permease and will react with β-galactosidase, if present, to produce a yellow color (see Fig. 5-47).

Because β-galactosidase is an inducible enzyme, it will be produced only in the presence of its inducer (the substrate β-lactose). To avoid false negative results, organisms being tested for the presence of β-galactosidase are cultured in a lactose rich medium (*e.g.* Kligler's iron agar or Triple sugar iron agar) prior to the test.

Many variations of this test are available. However, all employ o-nitrophenyl-β-D-galactopyranoside (ONPG) or p-nitrophenyl-β-D-galactopyranoside (PNPG) and all demonstrate essentially the same biochemistry. In the form of test illustrated here, ONPG tablets are dissolved in distilled water and a heavy inoculum of the organism being tested is added. A color change to yellow within a few hours is considered a positive result (see Fig. 5-48).

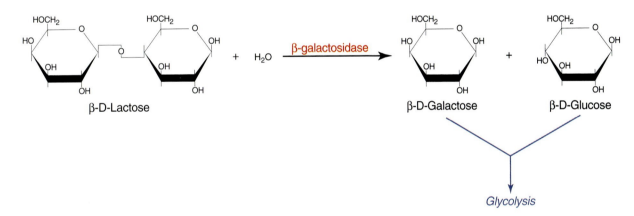

**Figure 5-46.** Hydrolysis of lactose by β-galactosidase.

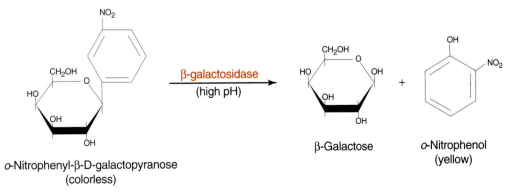

**Figure 5-47.** Conversion of ONPG to β-galactose and o-nitrophenol by β-galactosidase.

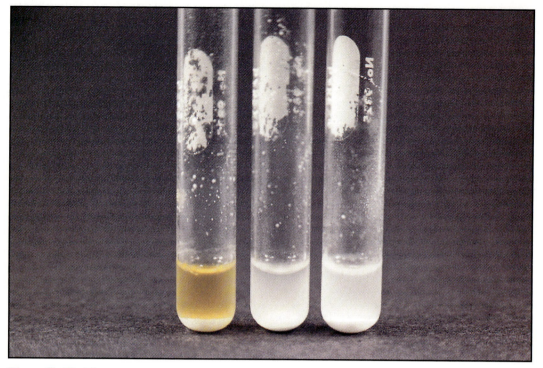

**Figure 5-48.** The ONPG test. *Escherichia coli* (+) on the left and *Proteus vulgaris* (−) on the right. An uninoculated control is in the middle.

*A Photographic Atlas for the Microbiology Laboratory*

# OXIDATION-FERMENTATION (O-F) MEDIUM

**Purpose**    This test is used to differentiate bacteria based on their ability to oxidize or ferment a specific sugar.

**Medical Applications**    The Oxidation-Fermentation test is used to separate pathogenic genera *Moraxella* (inert), *Pseudomonas* (O), *Alcaligenes* and *Bordetella* from members of the Enterobacteriaceae (F). *Staphylococcus* (F) and *Micrococcus* (usually O) can also be separated.

**Principle**    O-F test medium contains a high sugar to peptone ratio. This imbalance favors weak acid production and minimizes the production of ammonia (via deamination) and amines which neutralize the acids. The pH indicator used is bromthymol blue which detects even the very weak acids produced in oxidative metabolism. Bromthymol blue is yellow at pH 6.0, green at pH 7.1 and blue at pH 7.6.

O-F test medium is prepared to include any one of a variety of sugars. Typically, the medium used for testing enterics includes glucose, lactose, sucrose, maltose, mannitol or xylose. Staphylococcal O-F medium uses only glucose or mannitol.

Two tubes of the specific sugar medium are stabbed with the organism being tested. After inoculation, one tube is sealed with sterile mineral oil and the other is left unsealed. Mineral oil retards oxygen diffusion into the medium and thus promotes anaerobic growth. The unsealed tube allows aerobic growth.

Organisms able to ferment or ferment and oxidize the sugar will turn the sealed and unsealed media yellow. Organisms able only to oxidize the sugar will turn the unsealed medium yellow and leave the sealed medium green or blue. Organisms not able to metabolize the sugar will either produce no color change or will turn the medium blue. The results are summarized in the table in Figure 5-49 and shown in Fig. 5-50.

| SEALED MEDIA | UNSEALED MEDIA | INTERPRETATION |
|---|---|---|
| green or blue | green or blue | The organism performs no sugar metabolism; it is *nonsaccharolytic* |
| green or blue | yellow | The organism performs oxidative metabolism (O) |
| yellow | yellow | The organism performs fermentation (F) or fermentation and oxidation (O/F) |

**Figure 5-49.** Summary of O-F medium results.

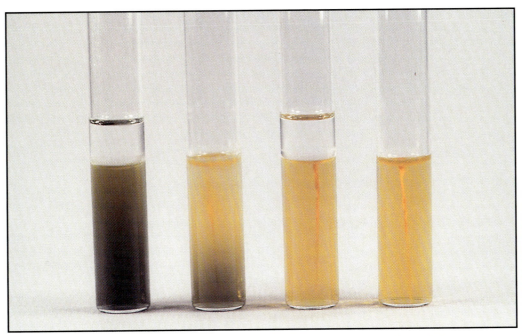

**Figure 5-50.** *Pseudomonas aeruginosa* (O) in the two tubes on the left and *Staphylococcus aureus* (O/F) in the two tubes on the right. Mineral oil in the first and third tubes prevents aerobic growth.

# OXIDASE TEST

**Purpose**    This test is used to identify bacteria containing the respiratory enzyme cytochrome *c*.

## Medical Applications

Oxidase tests may be used to identify the gonorrhea-causing bacteria, *Neisseria gonorrhoeae*. Opportunistic pathogens from the family Pseudomonadaceae (such as *Pseudomonas aeruginosa* which infects burn patients and may produce urinary infections) are usually oxidase (+) and can be differentiated from the members of the Enterobacteriaceae which are oxidase (−).

## Principle

Bacterial electron transport chains (ETCs), though quite diverse in specific composition, have the same basic purpose: to oxidize the NADH and FADH$_2$ produced in the Krebs cycle and glycolysis, and convert the energy released into bond energy in ATPs. ETCs consist of *flavoproteins, cytochromes* and *quinones*, all of which are capable of alternating between an oxidized and a reduced form. The oxidase test identifies the presence of a specific cytochrome, cytochrome *c* in the ETC.

Many aerobes and some facultative anaerobes utilize cytochrome *c* in their ETC. In the cell, cytochrome *c* receives electrons from cytochrome *b* and transfers them to cytochrome $a_1a_3$, which is then oxidized by oxygen, the final electron acceptor of aerobic respiration. A molecule of water is the result.

In the oxidase test, an artificial electron donor (either dimethyl-*p*-phenylenediamine or tetramethyl-*p*-phenylenediamine) is used to reduce cytochrome *c*. Oxidase (+) organisms will oxidize the colorless reagent and turn it to a dark blue color (see Fig. 5-51). If no cytochrome *c* is present, the reagent remains in its colorless reduced form, and the organism is identified as oxidase (−).

The oxidase test may be run by placing a few drops of the phenylenediamine solution on bacterial colonies (see Fig. 5-52) or by transferring a small amount of growth to a filter paper saturated with the reagent (see Fig. 5-53). In each case, a deep purple/blue color is considered a positive result. The oxidase reagent is unstable and will eventually oxidize in the absence of cytochrome *c*. A false positive can be avoided by reading the test within 60 seconds of reagent application.

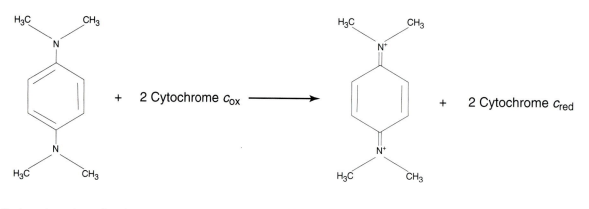

Tetramethyl-p-phenylenediamine$_{red}$
(colorless)

Tetramethyl-p-phenylenediamine$_{ox}$
(deep purple/blue)

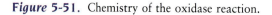

**Figure 5-51.** Chemistry of the oxidase reaction.

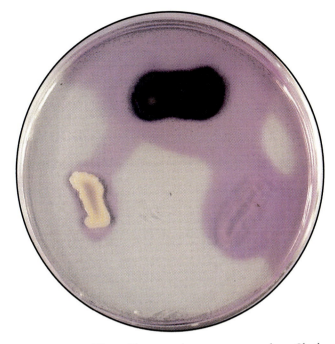

Figure 5-52. The oxidase test done on an agar plate. Clockwise from the top: *Pseudomonas aeruginosa* (+), *Clostridium sporogenes* (−), and *Staphylococcus aureus* (−). The reagent is very unstable and turns purple when exposed to air (as evidenced by the discoloration of the agar surface). The test should be read within 60 seconds to avoid false positive results.

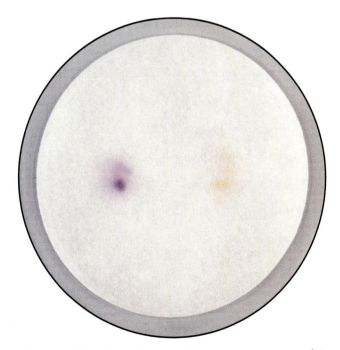

Figure 5-53. The oxidase test done on paper saturated in oxidase reagent. *Pseudomonas aeruginosa* (+) is on the left and *Staphylococcus aureus* (−) is on the right.

# PHENOL RED FERMENTATION BROTH

**Purpose**    This is a family of differential tests used to detect the ability of an organism to ferment various carbohydrates.

**Principle**    Fermentation is a metabolic process in which the final electron acceptor is an organic molecule.

Glucose fermentation typically begins with the production of pyruvic acid by glycolysis. (Some bacteria may alternatively use the pentose phosphate shunt or Entner-Doudoroff pathway, but they still produce pyruvic acid.) Various fermentation end products may be produced from pyruvic acid, including a variety of acids, $H_2$ or $CO_2$ gases, and alcohols. The specific end products depend on the organism and the substrate fermented (see Fig. 5-54).

Formulating this medium with different carbohydrates allows determination of a species' fermentation end products and its ability to enzymatically convert other sugars (*e.g.*

disaccharides) into glucose. The medium includes peptone, phenol red pH indicator (yellow at pH<6.8; red at pH>7.4) and the carbohydrate being tested. Acid production (A) will lower the pH and turn the phenol red yellow (see Fig. 5-55). An inverted tube (Durham tube) is included to trap any gas produced. A bubble is indicative of gas production (G).

The amino acids in peptone may be degraded by deamination which releases $NH_3$ into the medium. Accumulation of $NH_3$ raises the pH and turns the phenol red a fuchsia color. This alkaline reaction may be produced by bacteria that do not ferment the carbohydrate or by those that have exhausted the carbohydrate and have changed their metabolism to use another available resource. Because these latter organisms neutralize any acid they produced, it is important to read this test no later than 48 hours after inoculation to avoid missing their ability to produce acid.

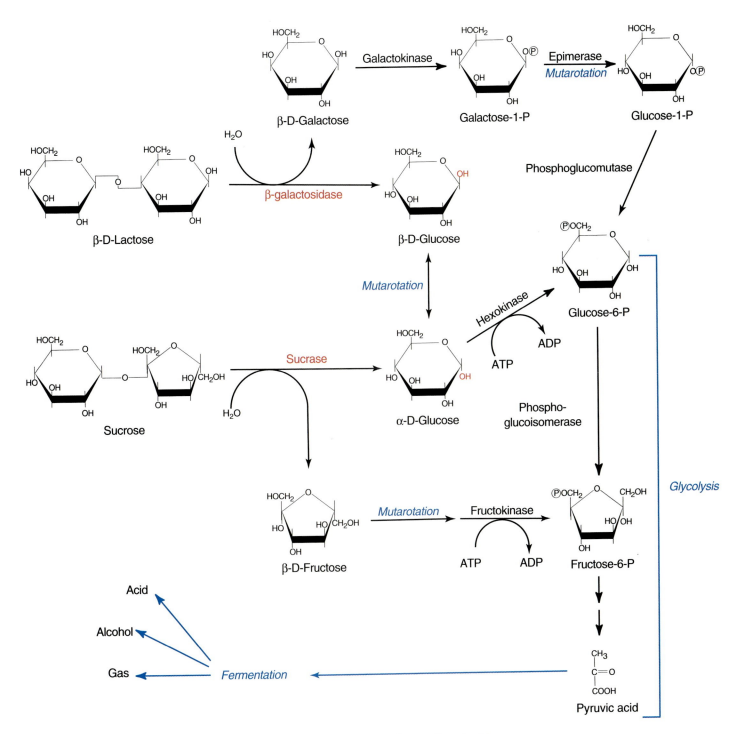

**Figure 5-54.** Fermentation of some disaccharides.

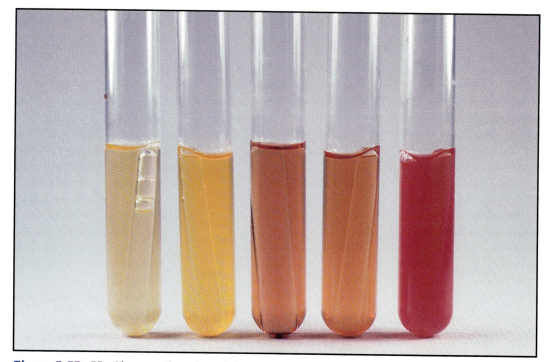

**Figure 5-55.** PR Glucose tubes, from left to right: *Escherichia coli* (A/G), *Staphylococcus aureus* (A/−), uninoculated control, *Micrococcus luteus* (−/−) and *Alcaligenes faecalis* (alk).

# PHENYLALANINE DEAMINASE AGAR

**Purpose** This medium is used to identify bacteria with the ability to deaminate phenylalanine to phenylpyruvic acid using the enzyme *phenylalanine deaminase*.

**Medical Applications** Phenylalanine deaminase medium is used to differentiate the genera *Morganella*, *Proteus* and *Providencia* (+) from other members of the Enterobacteriaceae (−). Bacteria in these genera are known to cause urinary tract infections and are capable of opportunistically causing septic lesions elsewhere in the body.

**Principle** Phenylalanine agar contains, among other things, yeast extract and DL-phenylalanine. Organisms containing the flavoprotein phenylalanine deaminase are capable of oxidatively converting the amino acid phenylalanine to phenylpyruvic acid and free ammonia by replacing the amine group with oxygen (see Fig. 5-56). Phenylpyruvic acid, which is colorless, may then be detected by adding an oxidizing reagent, such as 10% ferric chloride ($FeCl_3$). Ferric chloride reacts with phenylpyruvic acid and changes from yellow to green almost immediately (see Fig. 5-57). This indicates a positive result. Yellow color denotes a negative result (see Fig. 5-58). Since the green fades rapidly, this test should be read immediately to avoid a false negative result.

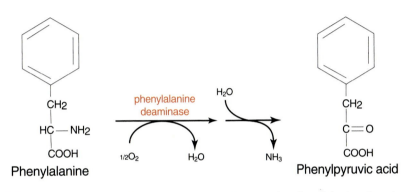

**Figure 5-56.** Deamination of the amino acid phenylalanine by phenylalanine deaminase.

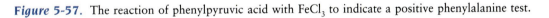

Phenylpyruvic Acid + $FeCl_3$ ⟶ Green Color

**Figure 5-57.** The reaction of phenylpyruvic acid with $FeCl_3$ to indicate a positive phenylalanine test.

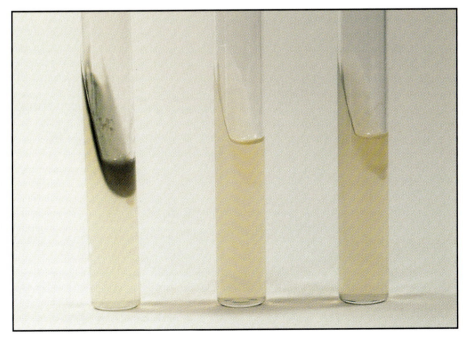

**Figure 5-58.** Phenylalanine deaminase tubes. *Proteus mirabilis* (+) on the left and *Escherichia coli* (−) on the right. An uninoculated control is in the center.

# SULFUR-INDOLE-MOTILITY (SIM) MEDIUM

**Purpose**    SIM (Sulfur-Indole-Motility) medium provides results for three different differential tests: sulfur reduction, indole production and motility. Sulfur reduction produces $H_2S$ either from the catabolism of the amino acid cysteine by the enzyme *cysteine desulfurase*, or by the reduction of thiosulfate in anaerobic respiration. Indole is produced when the amino acid tryptophan is deaminated by the enzyme *tryptophanase*.

**Medical Applications**    The indole test is one component of the IMViC test used in the identification of enteric bacteria. IMViC stands for *Indole, Methyl red, Voges-Proskauer,* and *Citrate* tests. Many enterics are pathogens: *Salmonella typhi* produces typhoid fever (indole −), *Klebsiella pneumoniae* causes pneumonia (indole −), and *Yersinia pestis* causes plague (indole −). *Escherichia coli*, an opportunistic pathogen, is usually indole (+).

The ability to reduce sulfur aids in the identification of many pathogenic and nonpathogenic bacteria. The most notable are genus *Salmonella* (enteric fevers, gastroenteritis and septicemia) usually (+) and *Francisella tularensis*, the highly infectious and virulent cause of tularemia, also (+). The test also differentiates the pathogens *Proteus mirabilis* (+) and *Proteus vulgaris* (+) from *Morganella morganii* (−) and *Providencia rettferi* (−).

**Principle**    The amino acid tryptophan can be used as a carbon and an energy source by bacteria possessing the enzyme *tryptophanase* (see Fig. 5-59). Pyruvic acid can then be used by the organism in the Krebs cycle or it can enter glycolysis and be used to synthesize other compounds the cell needs.

Indole medium typically includes peptone and necessary growth factors but is enriched with the amino acid tryptophan.

It may be either a solid medium such as SIM or a broth. After incubation, a small amount of test reagent (either Kovac's or Erlich's) is added to the medium. Both test reagents contain HCl and a form of Dimethylaminobenzaldehyde (DMABA) dissolved in amyl alcohol. Amyl alcohol is insoluble in water and will form a separate layer on top of the medium. In indole (+) organisms, the DMABA reacts with indole to produce a roseindole dye (see Fig. 5-60) which makes the alcohol layer a cherry red color (see Fig. 5-61). No color change is observed in the test reagent for indole (−) organisms.

Bacteria produce $H_2S$ either by putrefaction or by anaerobic respiration. In putrefaction sulfur containing amino acids (*e.g.* cysteine) are broken down by the enzyme *cysteine desulfurase* producing pyruvic acid, ammonia and $H_2S$ (see Fig. 5-62). The ammonia and hydrogen sulfide are excreted from the cell and the pyruvic acid is retained for energy production via the Krebs cycle. In the second process, a type of anaerobic respiration, inorganic sulfur (in this case thiosulfate) becomes the final electron acceptor of the electron transport chain (see Fig. 5-63).

The ingredients of SIM medium relevant to sulfur reduction are peptone, beef extract, sodium thiosulfate, and peptonized iron or ferrous sulfate. The peptone provides the cysteine, the sodium thiosulfate provides the reducible sulfur for anaerobic respiration, and the peptonized iron or ferrous sulfate serve as an indicator by reacting with any $H_2S$ produced and forming a black precipitate (see Fig. 5-64). The black precipitate is read as $H_2S$ (+). No change to black color is $H_2S$ (−). Results are shown in Figure 5-65.

A positive result for motility is seen as spreading from the stab line (see Fig. 5-41).

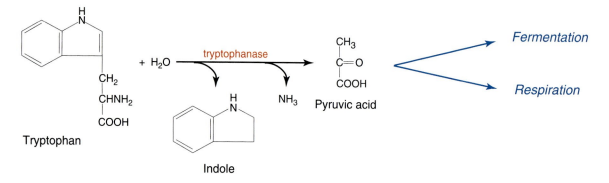

**Figure 5-59.** Tryptophan catabolism in indole (+) organisms.

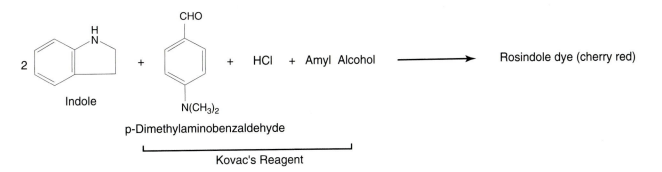

Figure 5-60. Indole reaction with Kovac's Reagent.

Figure 5-61. The indole test. SIM medium inoculated with *Morganella morganii* (indole +) on the left and *Enterobacter aerogenes* (indole −) on the right. An uninoculated control is in the center.

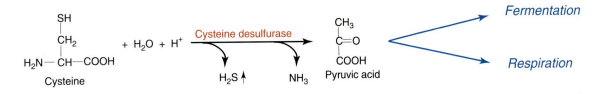

Figure 5-62. Putrefaction involving cysteine desulfurase.

$$3S_2O_3^= + 4H^+ + 4e^- \xrightarrow{\text{Thiosulfate reductase}} 2SO_3^= + 2H_2S\uparrow$$

Figure 5-63. Anaerobic respiration with thiosulfate as the final electron acceptor.

$$H_2S + FeSO_4 \longrightarrow H_2SO_4 + FeS\downarrow$$

Figure 5-64. Indicator reaction for $H_2S$, a colorless gas. The FeS produced is a black precipitate and indicates the presence of $H_2S$.

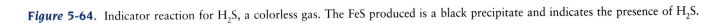

**Figure 5-65.** Sulfur reduction in SIM medium. On the left is *Proteus mirabilis* ($H_2S$ +) and *Pseudomonas fluorescens* ($H_2S$ −) on the right.

# STARCH AGAR

***Purpose***   This test is used to determine the ability of bacteria to enzymatically hydrolyze starch.

## Medical Application
*Corynebacterium diphtheriae* var. *gravis* (severe diphtheria) hydrolyzes starch and may be distinguished from *Corynebacterium diphtheriae* var. *intermedius* and *C. diphtheriae* var. *mitis* (less severe diphtheria) which are starch hydrolysis negative.

Other pathogens able to hydrolyze starch include *Vibrio cholerae* (cholera), *Clostridium perfringens* (gas gangrene), and *Bacillus anthracis* (anthrax). Some pathogens incapable of starch hydrolysis are *Pseudomonas aeruginosa* (opportunistic pathogen of skin and mucous membranes), *Clostridium botulinum* (botulism), and Group D streptococci (various infections).

## Principle
Starch is a polysaccharide made up of α-D-glucose subunits. It exists in two forms, linear (amylose) and branched (amylopectin), with the branched configuration being the predominant form in a mixture. The α-D-glucose molecules in both amylose and amylopectin are bonded by 1,4-α-glucosidic (acetal) linkages (see Fig. 5-66). The two forms differ in that the amylopectin contains polysaccharide side chains connected to approximately every 30th glucose in the main chain. These side chains are identical to the main chain except that the number 1 carbon of the first glucose in the side chain is bonded to carbon number 6 of the main chain glucose. The bond is, therefore, a 1,6-α-glucosidic linkage.

Since the intact polysaccharide is much too large to enter a cell, it must first be split into smaller fragments or individual glucose molecules to be of use to bacteria. Organisms that produce and secrete the extracellular enzymes *α-amylase* and *oligo-1,6-glucosidase* are able to hydrolyze starch by breaking the glucosidic linkages between sugar subunits. Although there usually are intermediate steps and additional enzymes utilized (depending on the organism), the overall reaction is the complete hydrolysis of the polysaccharide to its individual α-glucose subunits (see Fig. 5-66).

Starch agar is a simple plated medium of beef extract, soluble starch and agar. When organisms that produce α-amylase and oligo-1,6-glucosidase are grown on starch agar they hydrolyze the starch in the agar medium around the colonies. Because both the starch and its sugar subunits are soluble (clear) in the medium, the reagent iodine is used to detect the presence or absence of starch in the vicinity around the growth. Iodine reacts with starch and produces a blue color. Therefore, if iodine is poured onto a plate containing bacterial growth the agar will turn dark blue. Any clearing around the growth is the result of starch hydrolysis in that area and is considered a positive result.

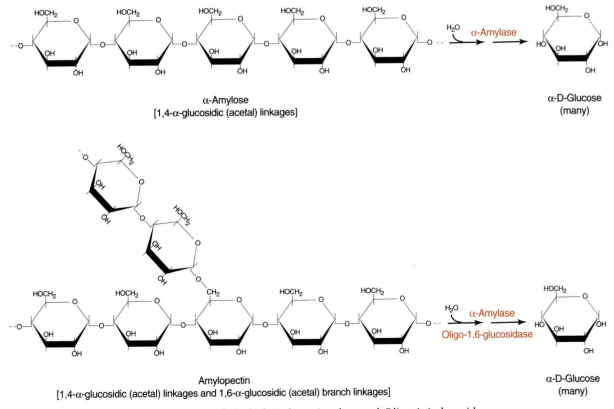

α-Amylose
[1,4-α-glucosidic (acetal) linkages]

α-D-Glucose
(many)

Amylopectin
[1,4-α-glucosidic (acetal) linkages and 1,6-α-glucosidic (acetal) branch linkages]

α-D-Glucose
(many)

***Figure 5-66.*** Starch hydrolysis by α-Amylase and Oligo-1,6-glucosidase.

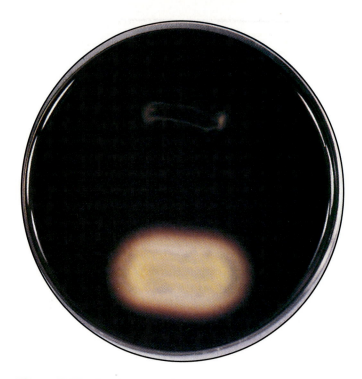

**Figure 5-67.** Starch agar with iodine added to detect amylase activity. *Escherichia coli* (−) is above and *Bacillus cereus* (+) is below.

*A Photographic Atlas for the Microbiology Laboratory*

**Purpose**    Tributyrin agar detects the exoenzyme lipase.

**Medical Applications**    The lipase test is not widely used in identifying pathogens, though the ability of *Staphylococcus aureus* to produce lipase accounts, in part, for its ability to spread through tissues. Lipolytic microorganisms are responsible for a category of food spoilage called "rancidity."

**Principle**    Triglycerides in the environment may serve as a carbon and energy source, but are too large to enter the cell. Some bacteria produce and secrete exoenzymes called *lipases* that hydrolyze triglycerides into a three carbon alcohol (glycerol) and three long chain fatty acids (see Fig. 5-68).

Glycerol may be converted into dihydroxyacetone phosphate, an intermediate of glycolysis. Fatty acids may be catabolized by a process called β-oxidation. Two carbon fragments from the fatty acid are combined with Coenzyme A to produce Acetyl CoA which may then be used in the Krebs cycle. Each Acetyl CoA produced by this process also yields one NADH and one $FADH_2$. Glycerol and fatty acids may also be used in anabolic pathways.

Tributyrin agar contains the triglyceride tributyrin and is initially opaque. Lipase (+) organisms will exhibit a clear zone around their growth as the tributyrin is digested (see Fig. 5-69).

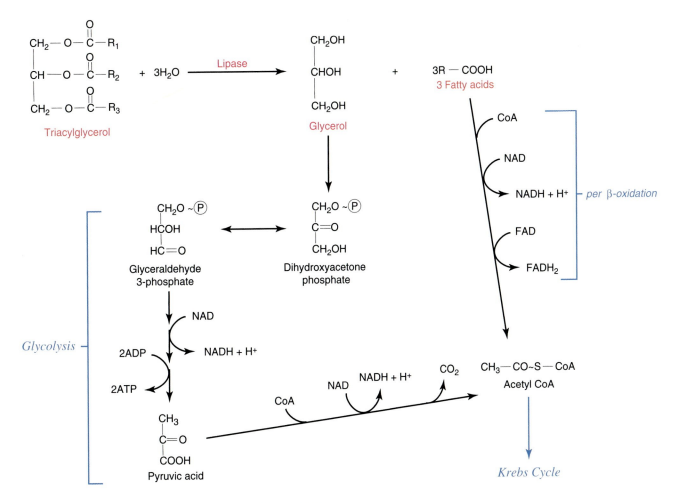

**Figure 5-68.** Lipid metabolism.

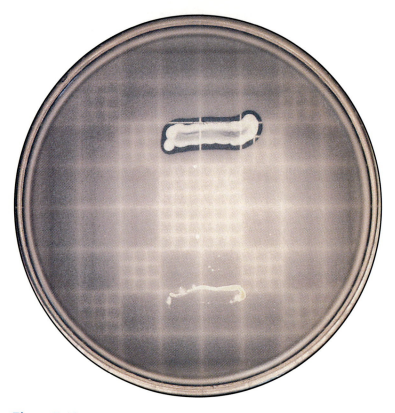

**Figure 5-69.** *Citrobacter amalonaticus* (+) above and *Erwinia amylovora* (−) below grown on tributyrin agar.

*A Photographic Atlas for the Microbiology Laboratory*

# TRIPLE SUGAR IRON (TSI) AGAR

**Purpose**   Triple sugar iron agar (TSI) differentiates bacteria based on their ability to ferment glucose, lactose and/or sucrose, and to reduce sulfur to hydrogen sulfide.

## Medical Applications

TSI is used to distinguish between members of the Enterobacteriaceae, all of which ferment glucose to an acid end product. Because they are so similar morphologically, a single medium providing a variety of results is of great use in their identification. Many members of this group are pathogenic, including *Escherichia coli* (opportunistic urinary infections), *Salmonella typhi* (typhoid fever), *Shigella* (bacillary dysentery), *Yersinia pestis* (plague), and *Klebsiella* (pneumonia).

## Principle

Triple sugar iron agar is the same as Kligler's iron agar with the exception of a third sugar (sucrose) added to TSI. For a description of the biochemistry of TSI, the reader is referred to the section on KIA.

The various reactions in TSI are the same as with KIA, with the exception that sucrose fermentation is added as a possible result (see Fig. 5-70 and Fig. 5-71).

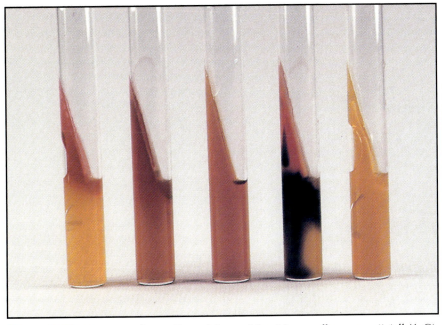

**Figure 5-70.** TSI agar slants. From left to right: *Morganella morganii* (alk/A,G), *Pseudomonas aeruginosa* (alk/NC), uninoculated control, *Proteus mirabilis* (alk/A,H₂S), and *Escherichia coli* (A/A,G).

| RESULTS | SYMBOL | INTERPRETATION |
|---|---|---|
| red/yellow | alk/A | glucose fermentation only<br>peptone catabolized |
| yellow/yellow | A/A | glucose and lactose and/or sucrose fermentation |
| red/red | alk/alk | no fermentation<br>peptone catabolized |
| red/no color change | alk/NC | no fermentation<br>peptone used aerobically |
| yellow/yellow with bubbles | A/A,G | glucose and lactose and/or sucrose fermentation<br>gas produced |
| red/yellow with bubbles | alk/A,G | glucose fermentation only<br>gas produced |
| red/yellow with bubbles and black precipitate | alk/A,G, $H_2S$ | glucose fermentation only<br>gas produced<br>$H_2S$ produced |
| red/yellow with black precipitate | alk/A, $H_2S$ | glucose fermentation only<br>$H_2S$ produced |
| yellow/yellow with black precipitate | A/A, $H_2S$ | glucose and lactose and/or sucrose fermentation<br>$H_2S$ produced |
| yellow/yellow with bubbles and black precipitate | A/A, G, $H_2S$ | glucose and lactose and/or sucrose fermentation<br>gas produced<br>$H_2S$ produced |
| no change/no change | NC/NC | no fermentation (organism is growing very slowly or not at all) |

*Figure 5-71.* Results, symbols and interpretations of TSI agar.

*A Photographic Atlas for the Microbiology Laboratory*

# UREASE BROTH AND AGAR

**Purpose**   This test is used to detect the presence of the enzyme *urease* which removes two molecules of ammonia from each molecule of urea.

**Medical Applications**   Members of the genus *Proteus* may be distinguished from other enteric bacteria by their fast urease activity. *P. mirabilis* is a major cause of human urinary tract infections. Species that are urease positive are also important ecologically as decomposers, as evidenced by cat litter box and diaper pail odors.

**Principle**   Hydrolysis of urea by urease occurs as shown in Figure 5-72. Ammonia may then be used as a nitrogen source during amination reactions to produce amino acids and nucleotides.

Urease medium contains essential nutrients for growth and may either be solid or liquid. In either case, it also contains urea $(NH_2)_2C=O$ and phenol red (a pH indicator). Accumulation of ammonia raises the pH. When the pH reaches 8.4, the phenol red turns from yellow to a pink. A pink color is a positive result.

With solid media, rate of urea hydrolysis may be determined by the degree of red color in the slant: score a +4 if the entire tube is red, +2 if only the butt remains yellow, and weak + if only the very top of the slant is red.

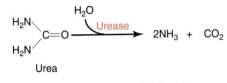

**Figure 5-72.** Urease biochemistry.

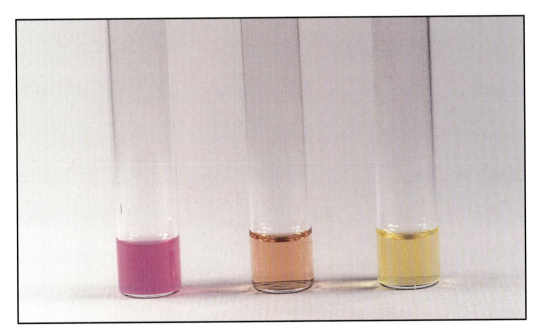

**Figure 5-73.** *Morganella morganii* (+) on the left and *Hafnia alvei* (−) on the right in urea broth. An uninoculated control is in the center.

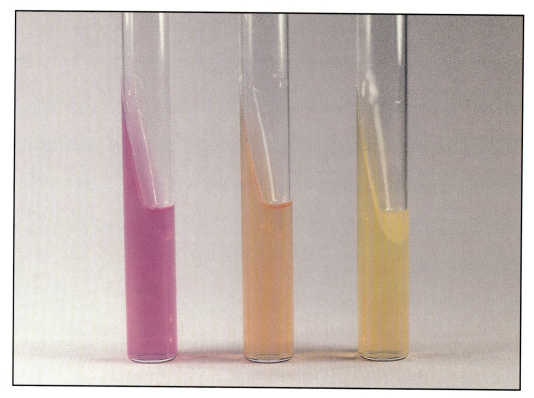

**Figure 5-74.** Urea agar tubes inoculated with *Morganella morganii* (+) on the left and *Hafnia alvei* (−) on the right. An uninoculated control is in the center.

*A Photographic Atlas for the Microbiology Laboratory*

# Quantitative Techniques

# VIABLE COUNT

**Purpose**  The viable count method is used to determine density of living cells in a sample. It may also be used to plot bacterial growth in a closed system by measuring population size at regular intervals.

**Medical Applications**  In research it is frequently necessary to have an accurate count of living (viable) cells in a culture. This procedure, done properly, can give quite accurate results.

**Principle**  The viable count involves plating a sample and counting the resulting colonies after incubation. It differs from the direct count in that the viable count is an indirect method (since the cells are not actually seen) and measures only living cells. As with direct counting, the viable count is performed on young, actively growing cultures.

A *serial dilution* (described in Figure 6-1) of the original sample is performed prior to plating. The serial dilution is used to reduce the number of organisms to a manageable concentration because only plates with between 30 and 300 colonies are considered *countable*. Plates that have fewer than 30 colonies have insufficient numbers to be reliable and plates with more than 300 are too crowded to count accurately (see Figs. 6-2 through 6-4). The dilution series also provides the frame of reference for calculating the concentration of the original sample.

Primarily for convenience and to maintain the viability of the organisms the preferred method of plating dilutions is the *spread plate technique*. Platings from each of several dilutions are performed. If growth dynamics of the population are being determined, dilutions and multiple plates are made at the appointed times.

When a diluted cell sample has been plated, the number of colonies produced can be used to calculate the original cell density of the sample using the following formula:

$$\text{original cell density} = \left(\frac{\text{\# colonies}}{\text{milliliters plated}}\right)\left(\frac{1}{\text{dilution factor}}\right)$$

For example if 120 colonies (representing 120 cells introduced onto the plate) are counted on a plate inoculated with 1 ml of a $10^{-5}$ dilution factor,

$$\text{original cell density} = \left(\frac{\text{\# colonies}}{\text{milliliters plated}}\right)\left(\frac{1}{\text{dilution factor}}\right)$$

$$\text{original cell density} = \left(\frac{120\ \text{cells}}{1\ \text{ml}}\right)\left(\frac{1}{10^{-5}}\right)$$

$$\text{original cell density} = 1.2 \times 10^7\ \text{cells/ml}$$

Because it is virtually impossible to know if the colonies produced on the plate originated with a single cell or a cluster of cells, the term *CFU (colony forming units)*/ml is conventionally used instead of *cells*/ml. The correct answer to the preceding problem would more correctly be $1.2 \times 10^7$ CFU/ml.

The advantages of viable count are that it gives a better representation of growth rate since it measures only living cells. The major disadvantage is that it is highly dependent upon accuracy and extremely careful technique. A careless measurement can yield grossly inaccurate results.

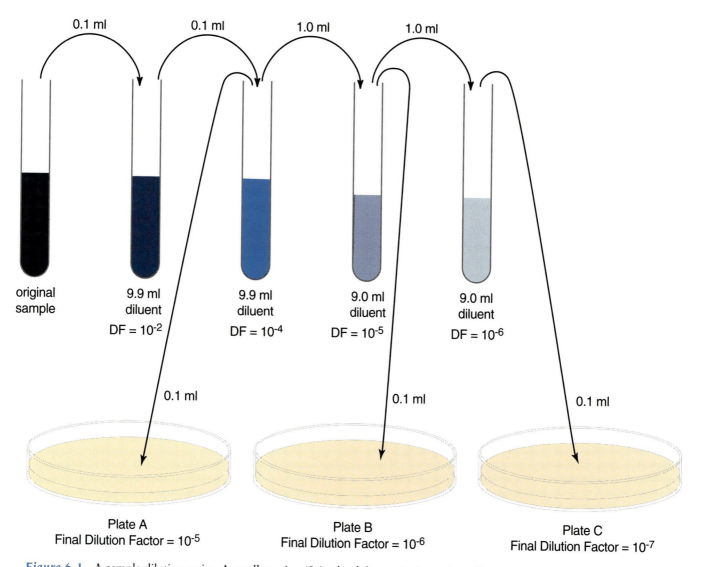

**Figure 6-1.** A sample dilution series. A small portion (0.1 ml) of the original sample is diluted in a larger volume (9.9 ml) of the first tube. This is followed by a transfers from the first tube into the second, the second into the third, *etc.* Each specific dilution is assigned a dilution factor. Dilution factor (DF) is determined by dividing the volume being transferred ($V_1$) by the total resulting volume ($V_2$). The dilution factor for 0.1 ml added to 9.9 ml would be $10^{-2}$, as illustrated below.

$$DF = \frac{V_1}{V_2} = \frac{0.1 \text{ ml}}{10.0 \text{ ml}} = 10^{-2}$$

To obtain the dilution factor of the plate multiply the DF of the solution being transferred by the volume transferred. Suppose, for example, 0.1 ml of a $10^{-4}$ solution is plated,

$$\text{Final DF} = (10^{-4})(0.1) = 10^{-5}$$

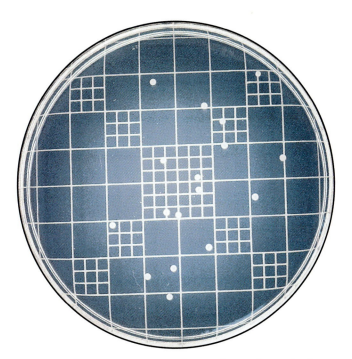

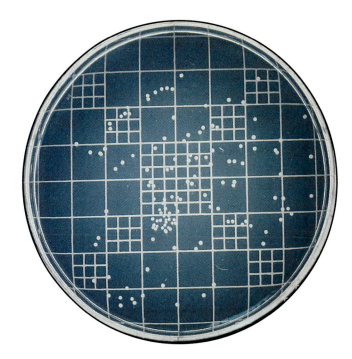

**Figure 6-2.** A viable count plate with too few colonies to provide a statistically reliable estimate of cell density. The plate is scored as TFTC ("too few to count").

**Figure 6-3.** A countable plate has between 30 and 300 colonies.

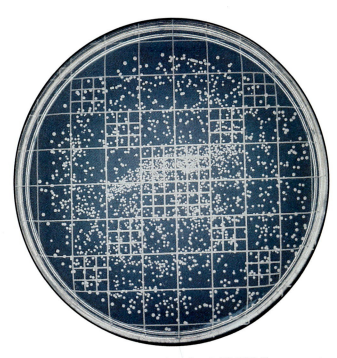

**Figure 6-4.** A viable count plate that is TMTC ("too many to count") because there are more than 300 colonies on it.

# DIRECT COUNT — PETROFF-HAUSSER COUNTING CHAMBER

**Purpose**    The direct count method is used to determine bacterial cell density in a sample. It may also be used to calculate an organism's generation time and growth rate if a series of measurements are made on a single population.

**Principle**    Counting all the bacteria, even in a small sample, would be virtually impossible. A direct count is performed using a device called a *Petroff-Hausser counting chamber* that allows counting of a small volume of diluted sample. This device, a specially modified microscope slide, contains a chamber or "well" of known depth (usually 0.02 mm) in the center with an etched grid on the bottom (see Fig. 6-5). The grid has an area of 1 mm² and consists of 25 large squares marked by double lines. Inside each of these larger squares is a grid of 16 small squares marked by single lines. Therefore the total number of squares in 1mm² is 16 × 25 or 400 squares. This grid of small squares is the area used for counting bacteria (see Fig. 6-6).

The volume above each small square is $5 \times 10^{-8}$ ml. Cell density is usually reported in cells/ml, so calculation of original cell density must extrapolate from the volume in the

chamber up to an entire milliliter. This is done by dividing by the volume of sample used for counting (*i.e.* the volume above one small square). Any dilution must also be taken into account. This is done by multiplying by the reciprocal of the dilution factor.

For example, if an average of 16 cells/small square is counted in a sample that has been diluted by $10^{-4}$, the cell density in the original sample is

$$\text{original cell density} = \frac{\text{cells/small square}}{\substack{\text{volume above} \\ \text{a small square}}} \times \frac{1}{\text{dilution factor}}$$

$$\text{original cell density} = \frac{16 \text{ cells/small square}}{5 \times 10^{-8} \text{ ml/small square}} \times \frac{1}{10^{-4}}$$

$$\text{original cell density} = 3.2 \times 10^{12} \text{ cells/ml}$$

The advantages of direct counting are that it is quick, easy to do and is relatively inexpensive. The major disadvantage is that both living and dead cells are counted.

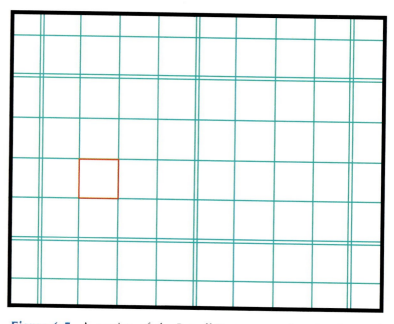

**Figure 6-5.** A portion of the Petroff-Hausser counting chamber grid with a small square highlighted in red. The volume above a small square is $5 \times 10^{-8}$ ml. (Each small square is 0.05mm × 0.05mm giving an area of $2.5 \times 10^{-3}$ mm². Since the well is 0.02mm deep, the volume above a small square is 0.02mm × $2.5 \times 10^{-3}$ mm² = $5 \times 10^{-5}$ mm³. Converting to cubic centimeters, $5 \times 10^{-5}$ mm³ × $10^{-3}$ cm³/mm³ = $5 \times 10^{-8}$ cm³. A cubic centimeter is the same as a milliliter, so the final units are: $5 \times 10^{-8}$ ml.)

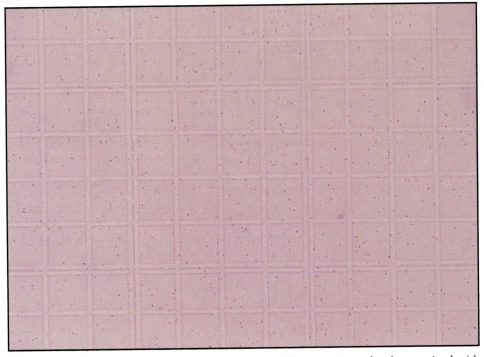

**Figure 6-6.** The Petroff-Hausser counting chamber. *Vibrio natriegens* has been stained with crystal violet and is visible on the grid.

# PLAQUE ASSAY
# FOR DETERMINATION OF PHAGE TITRE

**Purpose**     This technique is used to determine the concentration of viral particles in a sample.

**Principle**     Viruses that attack bacteria are called *bacteriophages* or simply *phages*. One way they do this, called the *lytic cycle* (see Fig. 9-2), is by attaching to the bacterial cell wall and injecting viral DNA into the bacterial cytoplasm. The viral genome then commands the cell to produce more viral DNA and viral proteins which are used for the assembly of more phages. Once assembly is complete, the cell lyses and releases the phages which attack other bacterial cells and begin the process all over again.

Lysis of bacterial cells growing on an agar plate and the subsequent attack of other cells in the immediate vicinity produces a clearing that can be viewed with the naked eye. These clearings are called *plaques*. Plaque assay uses this phenomenon as a means of calculating the phage concentration in a given sample. In this procedure, as in the viable count, a dilution series is used to reduce the number of particles to a manageable number. (For an explanation of dilution series, see Fig. 6-1.)

In contrast to the spread plate technique used in the viable count, plaque assay utilizes the *pour plate technique*. The pour plate technique adds a step to the conventional serial dilution in that the dilutions are not plated directly. Rather, they are added to a warm emulsion of dilute nutrient agar and young, actively growing bacteria. The mixture is then mixed gently, poured onto nutrient agar plates, allowed to cool, and incubated. The bacteria will create a bacterial lawn that, because of the agar overlay, will be immobile. The phages, on the other hand, will be able to diffuse through the soft agar and infect many cells in the immediate area. If the number of phages is significantly less than the number of bacteria and if the emulsion was mixed well, each plaque formed can be assumed to have originated from a single virus or *PFU* (*plaque forming unit*). As with the viable count, a countable plate should have between 30 and 300 PFU on it (see Figs. 6-7 through 6-9).

The plaques are counted and used to calculate the number of PFUs in the original sample using the following formula:

$$\text{original phage density} = \left(\frac{\text{\# plaques counted}}{\text{milliliters plated}}\right)\left(\frac{1}{\text{dilution factor}}\right)$$

Suppose that 150 plaques are counted on a plate which received 0.1 ml of a $10^{-4}$ dilution.

$$\text{original phage density} = \left(\frac{\text{\# plaques counted}}{\text{milliliters plated}}\right)\left(\frac{1}{\text{dilution factor}}\right)$$

$$\text{original phage density} = \left(\frac{150 \text{ plaques}}{0.1 \text{ ml}}\right)\left(\frac{1}{10^{-4}}\right)$$

$$\text{original phage density} = 1.5 \times 10^{7} \text{ PFU/ml}$$

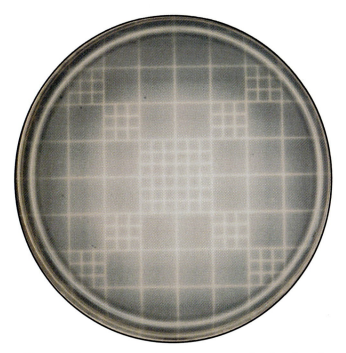

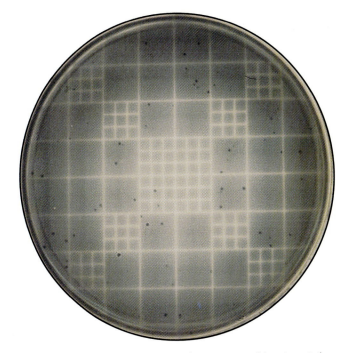

**Figure 6-7.** An uncountable plaque assay plate has less than 30 plaques and is scored as TFTC ("too few to count"). Less than 30 plaques is not a statistically sound sample size.

**Figure 6-8.** This plaque assay plate is countable, since it has between 30 and 300 plaques.

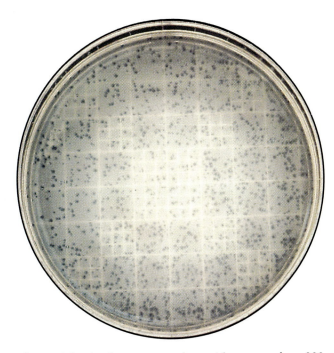

**Figure 6-9.** A plaque assay plate with greater than 300 plaques (TMTC). More than 300 plaques is difficult to count accurately.

# Medical, Food and Environmental Microbiology

## AMES TEST

**Purpose** This test is used to determine mutagenicity of chemical substances. Since a majority of mutagens are also carcinogens, the Ames test provides a rapid and inexpensive way of screening suspected carcinogens. (An assumption being made is that any substance that is highly mutagenic to the bacteria is also carcinogenic to higher animals.)

**Principle** *Salmonella typhimurium* has the enzymes necessary to manufacture all the amino acids it requires from its energy source. Certain strains of *S. typhimurium* (called histidine *auxotrophs*, as opposed to the original *prototrophs*) have mutated and lost their ability to synthesize histidine. Some of these are *frameshift mutants*. That is, they are missing one nucleotide in the sequence which would otherwise code for an enzyme necessary for histidine production. They are also missing the DNA repair mechanism which could correct the problem. Other strains are *substitution mutants*, in which one nucleotide in the histidine gene has been replaced, resulting in a faulty gene product. Histidine auxotrophs survive quite well, but only in an environment which provides the histidine they need.

The Ames test determines the ability of chemical agents to cause a reversal (*back mutation*) of these mutant conditions. When a small amount of the histidine auxotroph is plated onto *minimal (incomplete) medium* (which contains only glucose, a few salts, and trace amounts of histidine and biotin), the organism will exhibit faint growth, but will not develop into full-size colonies because they can't grow once

the histidine is exhausted. When a filter paper disc saturated with a suspected mutagen is placed in the middle of the minimal agar plate, the substance will diffuse outward into the medium. If it is mutagenic, it will cause back mutation in some cells (making them histidine prototrophs) which will grow into full-size colonies. (Note: histidine is initially included in the medium to allow the auxotrophs to grow for several generations and expose them to the effects of the mutagen.)

Several variations of the Ames test are possible. This example uses two minimal agar plates and two complete agar plates. Each plate is inoculated with histidine auxotroph. One plate each of minimal and complete agar then receives a filter paper disc saturated with the test substance. The second plate of minimal and complete agar then receives a filter paper disc saturated with a substance known to be nonmutagenic (in this case Dimethyl Sulfoxide, DMSO).

The purpose of the minimal/test plate is to determine mutagenicity of the test substance. Depending on the strain used, the type of mutation (either frameshift or base substitution) may also be determined. The minimal/DMSO plate serves as a control for the minimal/test plate by measuring the spontaneous back mutations. The purpose of the complete/test plate is to demonstrate toxicity of the substance by creation of a *zone of inhibition* around the disc. The complete/DMSO plate serves as the control for the zone of inhibition and growth on the complete/test plate (see Figs. 7-1 through 7-4).

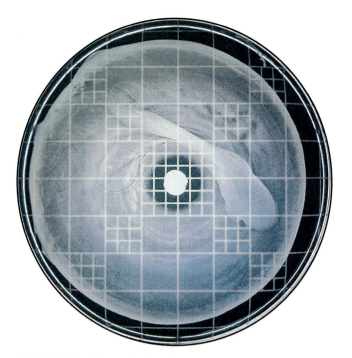

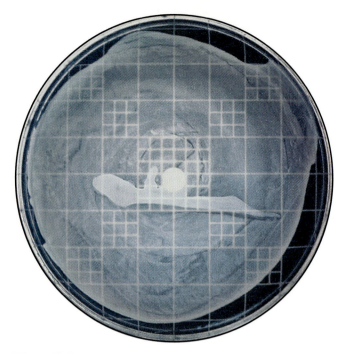

**Figure 7-1.** *Salmonella* histidine auxotroph grown on complete agar (containing histidine) and exposed to the test substance. The zone of inhibition around the paper disc indicates toxicity of the substance.

**Figure 7-2.** *Salmonella* histidine auxotroph grown on complete agar (containing histidine) and exposed to an inert substance (DMSO). This plate serves as a control for comparison to the plate in Figure 7-1.

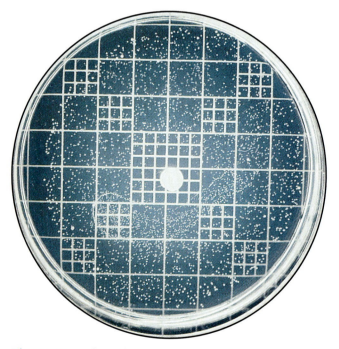

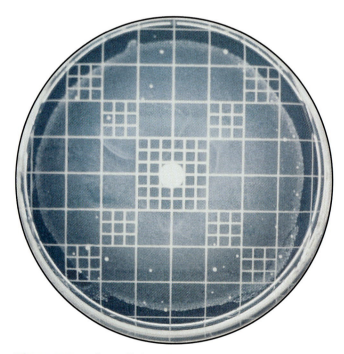

**Figure 7-3.** *Salmonella* histidine auxotroph grown on incomplete agar (histidine minimal agar) and exposed to a suspected mutagen. A zone of inhibition is visible on this plate, but is not as well-defined as on complete agar because the growth is not as dense.

**Figure 7-4.** *Salmonella* histidine auxotroph grown on incomplete agar (histidine minimal agar) and exposed to a known nonmutagen (DMSO). This plate is a control and demonstrates spontaneous mutation rate for comparison with the test substance (see Fig. 7-3). The faint background growth is due to small amount of histidine in the medium necessary to promote initial growth.

# ANTIBIOTIC SENSITIVITY—KIRBY-BAUER TEST

**Purpose**   Antibiotic sensitivity testing is used to determine the susceptibility of bacteria to various antibiotics.

**Medical Application**   This standardized testing method is used to measure the effectiveness of a variety of antibiotics on specific bacteria in order to prescribe the most suitable antibiotic therapy.

**Principle**   When an antibiotic impregnated paper disk is placed on a bacterial lawn growing in an agar plate it will create a clearing around the disk where the bacteria (if sensitive to the antibiotic) cannot grow. The size of this *zone of inhibition* depends upon the sensitivity of the bacteria to the specific antibiotic, the antibiotic's ability to diffuse from the disk through the agar, and very importantly, on the care taken in the preparation. Since the measurements are held to very exacting specifications this last point cannot be overemphasized.

The medium used for the Kirby-Bauer method is Mueller-Hinton agar. This agar is formulated to always have a pH between 7.2 and 7.4. It is poured into petri dishes to a depth of exactly 4 mm, and inoculated with a broth culture having a specific standardized turbidity, (McFarland No. 5 turbidity standard).

The antibiotic disks used for this test are also standardized for accuracy. The disks are dispensed onto the inoculated plate and incubated at 35°C. After exactly 18 hours incubation the plates are removed and the clear zones measured (see Figs. 7-5 and 7-6). The table in Figure 7-7 shows the standard interpretations of various antibiotic zones.

Normally, the zones on an agar plate will be distinct and separate halos around the antibiotic disk. On occasion, however, two zones will appear to join, producing an area of clearing between them extending beyond the perimeters of the otherwise circular zones (see Fig. 7-8). This is due to the *synergistic effect* of the two antibiotics. In other words, this is the area where the concentration of each antibiotic is too low to be effective by itself but, in combination with the other antibiotic, has sufficient strength to kill the bacteria.

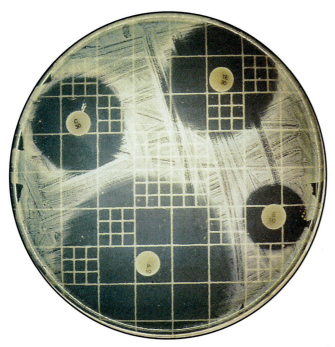

**Figure 7-5.** The Kirby-Bauer test illustrating the effect of (clockwise, from left) chloramphenicol (C30), tetracycline (TE30), streptomycin (S10), and penicillin (P10) on Gram-positive *Staphylococcus aureus*. The number on each disc indicates the amount of antibiotic in µg. As an example of how this test is used, the zone diameter for penicillin is approximately 50 mm. According to the Table in Figure 7-7, the minimum zone for susceptibility is ≥29 mm. Penicillin would be an effective antibiotic for use against this organism.

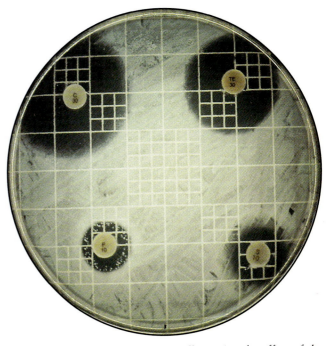

**Figure 7-6.** The Kirby-Bauer test illustrating the effect of the same antibiotics as in Figure 7-5 on Gram-negative *Escherichia coli*. Compare the zone produced by penicillin on this plate with that in Figure 7-5. Penicillin is generally ineffective against Gram-negative organisms. Note also the penicillin resistant colonies within the zone.

| Antimicrobial Agent | Code | Disc Potency | Zone Diameter Interpretive Standards (mm) | | |
|---|---|---|---|---|---|
| | | | Resistant | Intermediate | Susceptible |
| Chloramphenicol (for non-*Haemophilus* species) | C-30 | 30 µg | ≤12 | 13–17 | ≥18 |
| Penicillin (for staphylococci) | P-10 | 10 U | ≤28 | | ≥29 |
| Penicillin (for enterococci) | P-10 | 10 U | ≤14 | | ≥15 |
| Penicillin (for streptococci) | P-10 | 10 U | ≤19 | 20–27 | ≥28 |
| Streptomycin (for nonenterococci) | S-10 | 10 µg | ≤11 | 12–14 | ≥15 |
| Tetracycline (for most organisms) | Te-30 | 30 µg | ≤14 | 15–18 | ≥19 |
| Trimethoprim | TMP-5 | 5 µg | ≤10 | 11–15 | ≥16 |

**Figure 7-7.** Zone diameter interpretive chart for selected antibiotics based on data provided by the National Committee for Clinical Laboratory Standards (NCCLS). Permission to use portions of M100-S5 (*Performance Standards for Antimicrobial Susceptibility Testing; Fifth Informational Supplement*) has been granted by NCCLS. The interpretive data are valid only if the methodology in M2-A5 (*Performance Standards for Antimicrobial Disk Susceptibility Tests — Fifth Edition; Approved Standard*) is followed. NCCLS frequently updates the interpretive tables through new editions of the standard and supplements. Users should refer to the most recent editions. The current standard may be obtained from NCCLS, 940 West Valley Road, Suite 1400, Wayne, PA 19087.

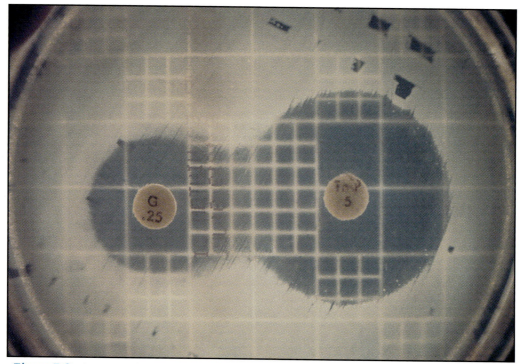

**Figure 7-8.** Synergism between the antibiotics Sulfisoxazole (G-.25) and Trimethoprim (TMP-5).

*A Photographic Atlas for the Microbiology Laboratory*

# MEMBRANE FILTER TECHNIQUE

**Purpose**    This technique is used to indicate the presence of coliform bacteria, especially *Escherichia coli*.

## Medical Application

Contaminated water has always been and remains today a serious public health concern. In this country, public drinking water is tested and treated daily to maintain *potability* and assure public safety. The eosin methylene blue (EMB) membrane filter technique is one test commonly used in combination with others to test for the presence of fecal *coliforms*.

Fecal contamination of drinking water or water used to wash food can lead to transmission of typhoid fever (*Salmonella typhi*), paratyphoid fever (*S. paratyphi*), bacillary dysentery (*Shigella dysenteriae*), and cholera (*Vibrio cholerae*). Viral diseases, such as poliomyelitis and infectious hepatitis (hepatitis type A), protozoan diseases such as amoebic dysentery (*Entamoeba histolytica*) and various diseases due to parasitic worms are also transmitted by fecally contaminated water.

## Principle

Fecal contamination is one of the most common pollutants in open water and, as such, has the potential of introducing some very serious pathogens to drinking water. The rarer and more pathogenic organisms found in water are usually in such low concentration that to test for them directly would be costly and time consuming. Instead, conventional procedures test for the presence of the much more prevalent coliform bacteria to indicate the possible presence of water borne pathogens.

*Escherichia coli* is the target organism of EMB agar because it grows in such abundance in the intestine and survives a long time in open water. Other members of the coliform group live not only in the intestine of animals but in the soil and on plants, so their presence in water does not necessarily indicate fecal contamination. It is assumed, therefore, that if *E. coli* is not present, there is no fecal contamination.

EMB agar contains peptone, lactose, and the dyes eosin Y and methylene blue. The purpose of the dyes is twofold: 1) they inhibit the growth of Gram-positive organisms and, 2) they react with each other under acidic conditions and form a dark purple complex which serves as an indicator of lactose fermentation.

All coliforms ferment lactose to produce acid and gas, but at different rates and in differing amounts. Because each species takes up a certain amount of the dye as it grows, the colonies take on characteristic colors on EMB agar. It is this differential color change which allows for separation of the various coliforms. For example, *Enterobacter aerogenes*, which produces a little acid, turns pink and mucoid on EMB agar; *E. coli*, which produces a lot of acid, turns dark purple with a green metallic sheen on the surface of the colonies. This dark purple color with the characteristic green metallic sheen is a positive result for EMB. Absence of this color is a negative result (see Figs. 2-5 & 2-6)

In the EMB membrane filter technique a water sample of known volume suspected of containing contaminants is drawn through a sterile membrane filter. The filter is a special micropore membrane designed to allow water to pass but trap any microorganisms larger than 0.45µm. It is then placed onto an EMB plate. After 24 hours incubation at 35°C the plate is removed and the dark purple or green colonies counted (see Fig. 7-9). It is desirable to filter enough water to yield between 20 and 200 colonies, and samples of especially polluted water may be diluted to fall within this range. The number of colonies counted are then recorded as number of coliform colonies/100ml using the following equation:

$$\frac{\text{coliform colonies}}{100 \text{ ml}} = \frac{\text{coliform colonies counted} \times 100\text{ml}}{\text{ml sample filtered}}$$

To be *potable*, water must contain less than 1 coliform/100ml. For reliability this test can be performed several times and an average taken of the counts.

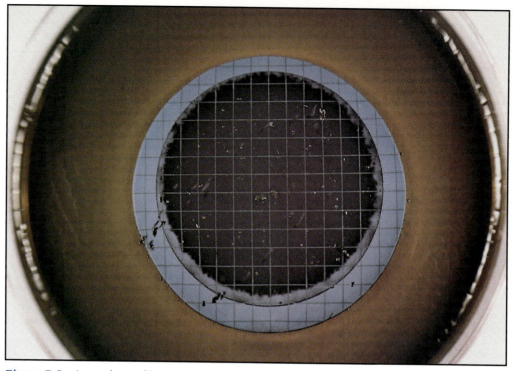

*Figure 7-9.* A membrane filter on eosin methylene blue (EMB) medium. The dark growth is characteristic of coliform bacteria and is indicative of water contamination. Typically, potable water has less than one coliform per 100 ml of water tested.

*A Photographic Atlas for the Microbiology Laboratory*

# METHYLENE BLUE REDUCTASE TEST

**Purpose**    The methylene blue reductase test can be used to separate organisms based on their ability to reduce methylene blue. It also, as in this example, can be used to measure the quality of raw milk.

**Medical Application**    This test is helpful in differentiating enterococci from other members of the genus *Streptococcus*. It also tests for the presence of coliforms in raw milk.

**Principle**    Methylene blue is a dye which is blue when oxidized and colorless when reduced. It can be reduced by a *reductase* enzyme either aerobically or anaerobically. Aerobically methylene blue can be substituted for any substrate in the electron transport system. It is reduced by cytochromes to leuco-methylene blue which is colorless, but is spontaneously oxidized again by oxygen. In this reaction hydrogen peroxide is produced but there is no color change.

Anaerobically, methylene blue is reduced by nucleotides such as nicotinamide adenine dinucleotide (NADH) or diphosphopyridine nucleotide (DPNH) to leuco-methylene blue and, in the absence of an oxidizing substance (*i.e.* oxygen) remains colorless.

The reduction of methylene blue may be used as an indicator of milk quality. A dilute methylene blue solution is added to a sterilized test tube containing raw milk. The tube is then tightly sealed and incubated in a 35°C water bath. The time it takes the milk to turn from blue to white (due to methylene blue reduction) is a qualitative indicator of the number of microorganisms living in the milk. Good quality milk takes greater than 6 hours to convert the methylene blue. The sample used in Figure 7-10 was purchased from a local health food store and, we are pleased to say, took approximately 20 hours to change from blue to white.

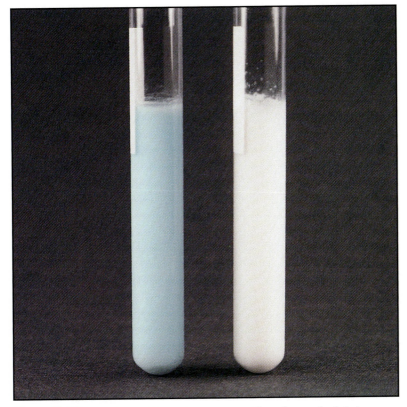

**Figure 7-10.** Methylene blue reductase test. The tube on the left is a control to illustrate the original color of the medium. The tube on the right indicates bacterial reduction of methylene blue after 20 hours. The speed of reduction is related to the concentration of microbes present.

# SNYDER TEST

**Purpose** The Snyder test is used to detect the presence of *Lactobacillus* in saliva.

**Medical Application** Tooth decay is a phenomenon that practically everyone is familiar with. Although several microorganisms are involved in the process, only species in the genus *Lactobacillus* seem to be capable of lowering the pH enough to dissolve tooth enamel. The Snyder test measures dental *caries* susceptibility by detecting the presence of Lactobacilli in saliva.

**Principle** Snyder Test medium is designed to favor growth of Lactobacilli and discourage growth of most other species. This is accomplished by adjusting the pH of the medium to 4.8 and by adding glucose, a carbohydrate easily fermented by the bacteria. Lactobacilli thrive in the low pH and ferment the glucose, producing more acid which reduces the pH even more. The medium includes the pH indicator, bromcresol green, which is green at pH 4.8 and above, and yellow below pH 4.8.

The medium is autoclaved for sterilization, cooled to just over 40 degrees and maintained in a warm water bath until needed. The still molten agar is then inoculated with 0.2 ml of saliva, mixed well, incubated, and allowed to grow for up to 72 hours. The agar tubes are checked at 24 hour intervals for any change in color. Yellow color indicates that fermentation has taken place and is a positive result (see Fig. 7-11). High susceptibility to dental caries is indicated if the medium turns yellow within 24 hours. Moderate and slight susceptibility are indicated by a change within 48 and 72 hours, respectively. No change within 72 hours is considered a negative result.

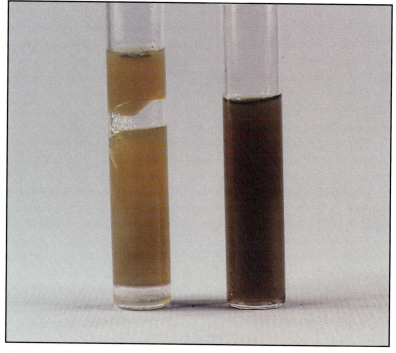

**Figure 7-11.** The Snyder test for dental caries susceptibility. A positive result is on the left, a negative on the right.

# Host Defenses, Immunology and Serology

## DIFFERENTIAL BLOOD CELL COUNT

**Purpose**   A differential blood cell count is done to determine approximate amounts of the various *leukocytes* (white blood cells). An excess or a deficiency of all or a particular group is indicative of certain disease states.

**Medical Applications**   White blood cells (WBCs) are divided into two groups: *granulocytes* (which have prominent cytoplasmic granules) and *agranulocytes* (which lack these granules). There are three basic classes of *polymorphonuclear granulocytes* or *PMNs*: *neutrophils, basophils,* and *eosinophils.* The two classes of agranulocytes are *monocytes* and *lymphocytes.*

   Neutrophils (see Figs. 8-1 through 8-3) are the most abundant WBC in blood. They leave the blood and enter tissues to phagocytize foreign material. An increase in neutrophils is indicative of a systemic bacterial infection. Mature neutrophils are sometimes referred to as *segs* because of their segmented nucleus, usually in two to five lobes. Immature neutrophils lack this segmentation and are referred to as *bands* (see Fig. 8-4). This distinction is useful, since a patient with an active infection will be producing more neutrophils, so a higher percentage will be of the band (immature) type. Neutrophils are 12-15µm in diameter.

   Eosinophils are phagocytic, and their numbers increase during allergic reactions and parasitic infections (see Fig. 8-5). They are about twice the size of an RBC (12-15µm) and generally have 2 lobes in their nucleus.

   Basophils (see Fig. 8-6) are the least abundant WBCs in normal blood. They are structurally similar to tissue mast cells and produce some of the same chemicals (histamine and heparin), but are derived from different stem cells in bone marrow. They are about twice the size of RBCs (12-15µm) and have two lobes in their nucleus or an unlobed nucleus.

   Agranulocytes include monocytes and lymphocytes. Monocytes (Fig. 8-7) are the blood form of macrophages. They are the largest of leukocytes, being two to three times the size of RBCs (12-20µm). Their nucleus is horseshoe shaped. Lymphocytes (Fig. 8-8) are cells of the immune system. Two functional types of lymphocytes are the T-cell, involved in cell-mediated immunity, and the B-cell, which

produces antibodies when activated. Lymphocytes are approximately the same size as RBCs or up to twice their size. The nucleus is usually spherical and takes up most of the cell.

**Principle**   A sample of blood is observed under the microscope and at least 100 white blood cells are counted and tallied. Approximate normal percentages for each leukocyte are as follows: 55-65% neutrophils (mostly segs), 25-33% lymphocytes, 3-7% monocytes, 1-3% eosinophils and 0.5-1% basophils.

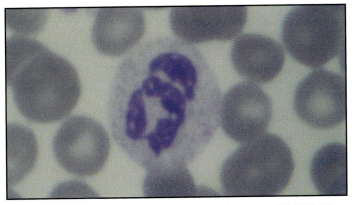

*Figure 8-1.* A segmented neutrophil (Wright's stain, X2376). Note its size relative to the RBCs and the lobed nucleus.

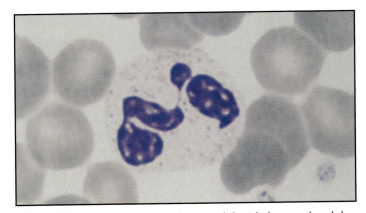

*Figure 8-2.* Another segmented neutrophil with four nuclear lobes (Wright's stain, X2640).

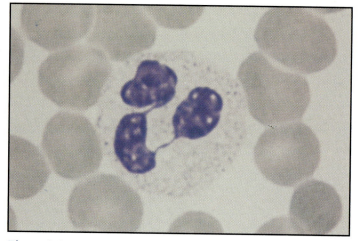

*Figure 8-3.* Another segmented neutrophil with three nuclear lobes (Wright's stain, X2640).

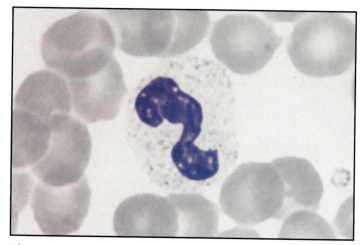

*Figure 8-4.* A band neutrophil (Wright's stain, X2640). Note the lack of nuclear segmentation.

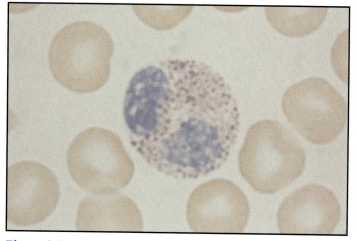

*Figure 8-5.* An eosinophil (Giemsa stain, X2640). Note the cell's size relative to the RBCs and its red cytoplasmic granules.

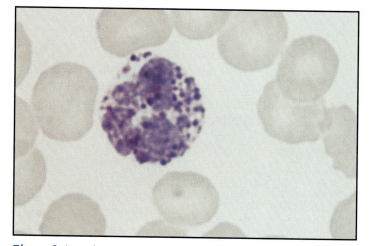

*Figure 8-6.* A basophil (Giemsa stain, X2640). Note the abundant purple cytoplasmic granules that partially obscure the nucleus.

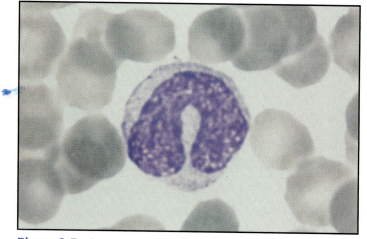

*Figure 8-7.* A monocyte (Wright's stain, X2640). Note its large size and horseshoe-shaped nucleus.

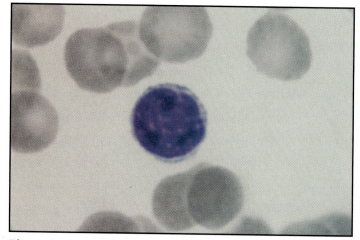

*Figure 8-8.* A lymphocyte (Wright's stain, X2640). Note its size compared to the RBCs in the field and the size of its nucleus relative to cell size. A small percentage of lymphocytes attain a size of up to 18μm.

Cells [...] system and immune system are involved in defense of the body and are found scattered throughout the body. Examples are shown in the following illustrations.

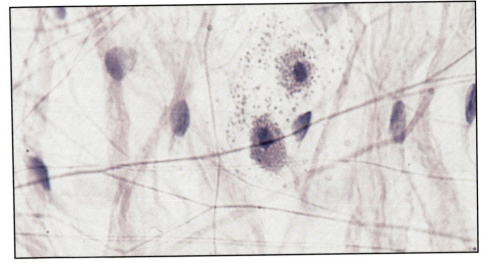

**Figure 8-9.** Two mast cells of loose connective tissue (X528). The cytoplasmic granules contain histamine and other chemicals involved in inflammation which are released by *degranulation* due to tissue damage. The mast cells may also be coated with IgE antibodies which cause degranulation when they bind antigen—as in Type I hypersensitivity (allergic) reactions.

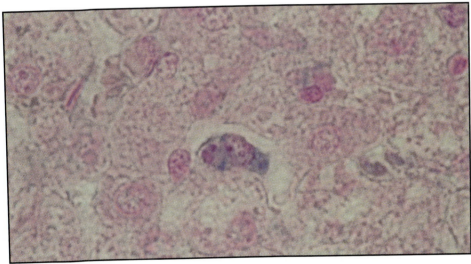

**Figure 8-10.** A section of the liver showing a Kuppfer cell (X1600). This example of a fixed macrophage has ingested a dye. Fixed macrophages may also be found in the lungs.

**Figure 8-11.** A section through a lymph node (X104). The large spherical purple objects are masses of lymphocytes called *lymph nodules* or *lymph follicles*. As lymph passes through the sinuses (channels) within the node, antigens may contact a lymphocyte with the ability to produce antibodies against it. This provides the stimulus for cloning of the lymphocyte and conversion of some clones into antibody-secreting plasma cells. Other clones become memory cells and reside in lymphatic tissue around the body. Macrophages are also found in the sinuses.

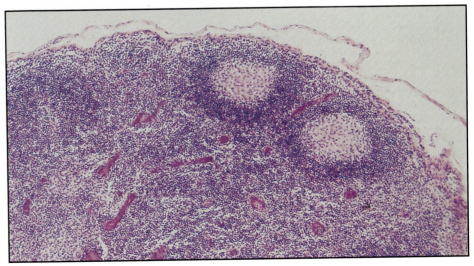

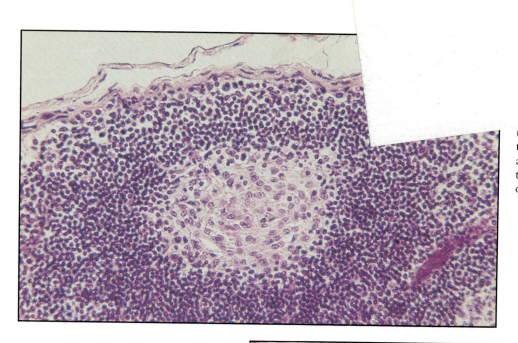

A single lymph nodule (X264). The germinal center (lighter, central portion) is occupied by lymphoblasts and medium-sized lymphocytes. More mature lymphocytes are found in the darker, outer portion of the nodule.

*Figure* 8-13. The germinal center of a lymph nodule (X792). Medium-sized lymphocytes are abundant in the lighter region, whereas mature lymphocytes predominate in the darker region. Their differences in size and nuclear staining are apparent.

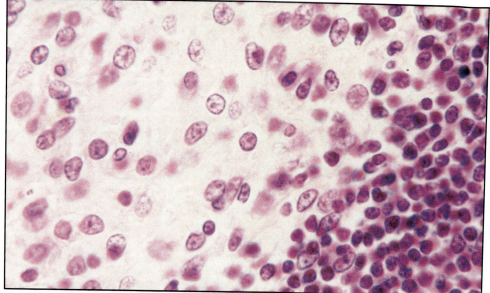

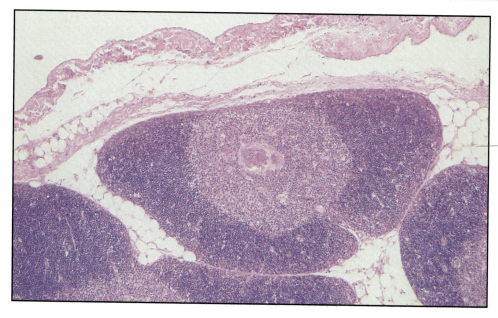

*Figure* 8-14. A section through the thymus (X104). One lobule (composed of lymphocytes) with its thymic corpuscle is shown. The thymus is the site of T-lymphocyte maturation.

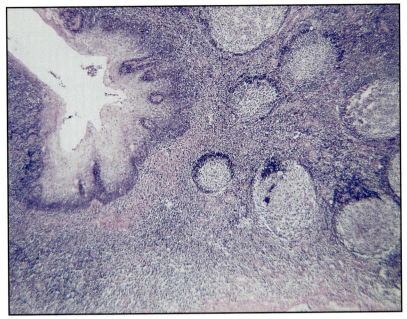

**Figure 8-15.** A section through a palatine tonsil (X58). Palatine tonsils are found on either side of the opening of the oral cavity to the throat. Two other sets of tonsils are also found in the throat: the pharyngeal tonsils (adenoids) are posterior to the nasal cavity and the lingual tonsils are in the base of the tongue. Tonsils act much like lymph nodes in that they are composed of lymph nodules (evident in the right half of the field) that contact fluids passing through them. A tonsilar crypt is visible on the upper left of the field.

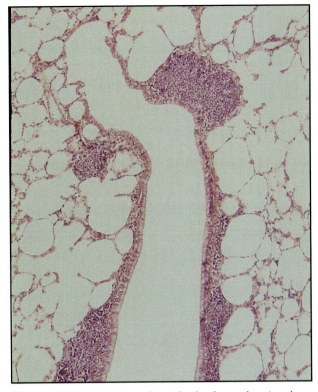

**Figure 8-16.** A section through the lung showing lymphatic tissue in the wall of a bronchiole (X100).

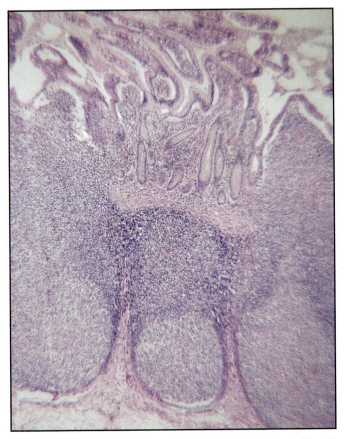

**Figure 8-17.** A section through the ileum showing a Peyer's patch of lymph nodules (X50). Peyer's patches may consist of 10 to 70 nodules separated from the intestinal lumen only by a single layer of epithelial cells.

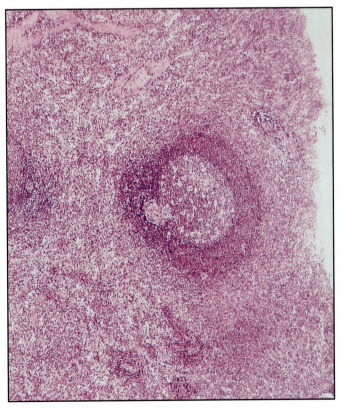

**Figure 8-18.** A section through a spleen (X80). In addition to being a blood filter and reservoir, the spleen contains lymph nodules referred to as *white pulp*. Lymphocytes of white pulp respond to antigens in the blood. The portion of the spleen devoted to blood-vascular functions is referred to as *red pulp*. A single nodule with its central artery is shown surrounded by the red pulp.

# PRECIPITATION REACTIONS

**Purpose**   Precipitation reactions may be used to detect either the presence of antigen or antibody in a sample.

**Medical Applications**   Precipitation tests have mostly been replaced by more sensitive serological techniques for diagnosis. Double-gel immunodiffusion may be used in identification of antibodies formed in autoimmune diseases.

**Principle**   *Soluble antigens* may combine with *homologous antibodies* to produce a visible *precipitate*. Precipitate formation thus serves as evidence of antigen-antibody reaction and is considered a positive result.

Precipitation is produced because each antibody has (at least) two antigen binding sites and many antigens have multiple sites for antibody binding. This results in the formation of a complex lattice of antibodies and antigens and produces the visible precipitate—a positive result. As shown in Figure 8-19, if either antibody or antigen is found in too high a concentration relative to the other, no visible precipitate will be formed even though both are present. *Optimum proportions* of antibody and antigen are necessary for precipitate formation and occur in the *zone of equivalence*.

Several styles of precipitation tests are used. The precipitin ring test is performed in a small test tube or capillary tube. Antiserum is placed in the bottom of the tube. The sample with the suspected antigen is layered on the surface of the antiserum in such a way that the two solutions have a sharp interface. As the two fluids diffuse into each other, precipitation occurs where optimum proportions of antibody and antigen are found (see Fig. 8-20). This test may also be run to test for antibody in a sample.

Double-gel immunodiffusion tests are used to check samples for identical, related or unrelated antigens. Wells are formed in a gel and a mixture of antibodies is placed in the center well. Samples with unknown antigens are placed in the surrounding wells (see Fig. 8-21). As antigens and antibodies diffuse radially from their respective wells, precipitation lines occur where optimal proportions are found. The precipitation line pattern is indicative of antigen relatedness: a single, smooth curved line indicates the two antigens in neighboring wells are identical ("identity"); a line with a spur indicates the antigens are related, but not identical ("partial identity"); two spurs indicate unrelated antigens ("nonidentity").

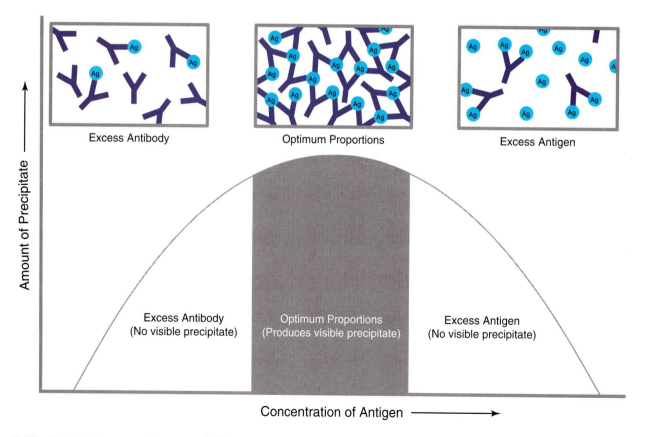

**Figure 8-19.** Precipitation occurs between soluble antigens and homologous antibodies where they are found in optimal proportions to produce a cross-linked lattice. Excess antigen or excess antibody prevent substantial cross-linking, so no lattice is formed and no visible precipitate is seen—even though both antigen and antibody are present.

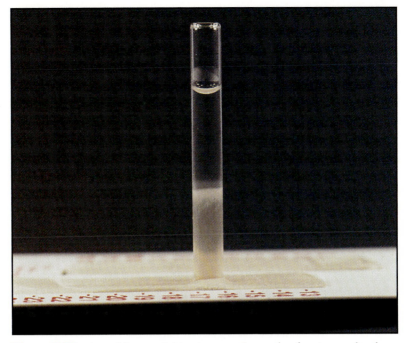

**Figure 8-20.** A positive precipitin ring test. A sample of antiserum has been layered over an antigen solution. The white precipitation ring formed at the site of optimal proportions of antibodies and antigens.

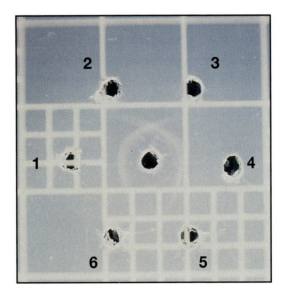

**Figure 8-21.** Double-gel immunodiffusion with antibodies in the center well and antigens in the outer wells. Precipitation lines formed between the central well and wells 1 and 2 illustrate identity of the antigens in wells 1 and 2. Identity is also illustrated by the antigens in wells 3 and 4. Nonidentity is illustrated by the lines formed between the central well and wells 2 and 3. No reaction occurred between the antibodies and antigens in wells 5 and 6.

# AGGLUTINATION REACTIONS

**Purpose**    Agglutination reactions may be used to detect either the presence of antigen or antibody in a sample.

**Medical Applications**    Direct agglutination reactions may be used to diagnose some diseases, determine if a patient has been exposed to a particular pathogen, and are involved in blood typing. Indirect agglutination is used in some pregnancy tests as well as in disease diagnosis.

**Principle**    Particulate antigens (such as whole cells) may combine with homologous antibodies to form visible clumps called *agglutinates*. *Agglutination* thus serves as evidence of antigen-antibody reaction and is considered a positive result.

There are many variations of agglutination tests (see Fig. 8-22). *Direct agglutination* relies on the combination of antibodies and naturally particulate antigens. *Indirect agglutination* relies on artificially constructed systems in which agglutination will occur. These involve coating particles (such as RBCs or latex microspheres) with either antibody or antigen, depending on what is being looked for in the sample. Addition of the appropriate antigen or antibody will then result in clumping of the artificially constructed particles.

Slide agglutination (see Figure 8-23) is an example of a direct agglutination test. Samples of antigen and antiserum are mixed on a microscope slide and allowed to react. Visible aggregates indicate a positive result.

*Hemagglutination* is a general term applied to any agglutination test in which clumping of red blood cells indicates a positive reaction. Blood tests as well as a number of indirect diagnostic serological tests are hemagglutinations.

The most common form of blood typing detects the presence of A and B antigens on the surface of red blood cells. An individual with type A blood has RBCs with the A antigen and produces anti-B antibodies. Likewise, an individual with type B blood has RBCs with the B antigen and produces anti-A antibodies. People with type AB blood have *both* A and B antigens on their RBCs and lack anti-A and anti-B antibodies. Type O individuals lack A and B antigens but produce *both* anti-A and anti-B antibodies.

ABO blood type is determined by adding a patient's blood to anti-A and anti-B antiserum and observing any signs of agglutination (see Fig. 8-24). Agglutination with anti-A antiserum indicates the presence of the A antigen and type A blood. Agglutination with anti-B antiserum indicates the presence

of the B antigen and type B blood. If both agglutinate, the individual has type AB blood; lack of agglutination occurs in individuals with type O blood.

A similar test is used to determine the presence or absence of the *Rh factor* (antigen). If clumping of the patient's blood occurs when mixed with anti-Rh (anti-D) antiserum, the patient is Rh positive (see Fig. 8-25).

Indirect hemagglutination may be used to detect the presence of either antigens or antibodies in a sample. In the example shown in Figure 8-26, sheep RBCs coated with *Treponema pallidum* (the causative agent of syphilis) antigen represent the particulate antigen. When added to antiserum containing anti-*Treponema pallidum* antibodies, agglutination occurs.

Direct agglutination occurs with naturally particulate antigens. Either antigen or antibody may be detected in a sample using this style of agglutination test.

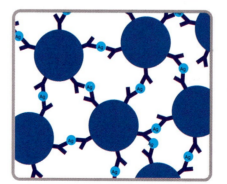

Detection of antibody in a sample by indirect agglutination. A test solution is prepared by artificially attaching homologous antigen (blue) to a particle (red) such as red blood cells or latex beads and mixing with the sample suspected of containing the antibody (purple).

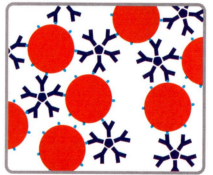

Another style of indirect agglutination detects antigen (light blue) in a sample. Antibodies (purple) are artificially attached to particles (dark blue).

*Figure 8-22.* Direct and indirect agglutination tests.

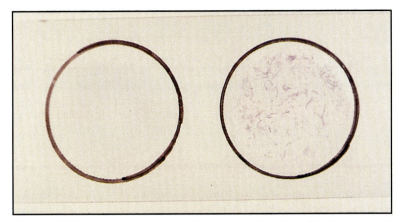

**Figure 8-23.** A positive slide agglutination of *Salmonella* H antigen is shown on the right. Anti-H antiserum mixed with *Salmonella* O antigen is shown on the left. Serological variation of H (flagellar) and O (somatic lipopolysaccharide) antigens is an important diagnostic feature of *Salmonella* serovars.

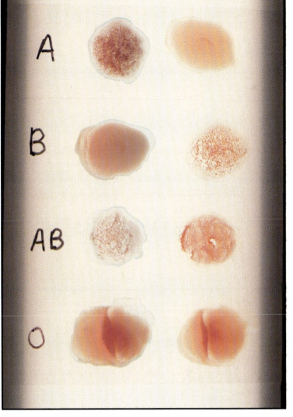

**Figure 8-24.** Blood typing relies on agglutination of RBCs by Anti-A or Anti-B antisera. The blood types are as shown.

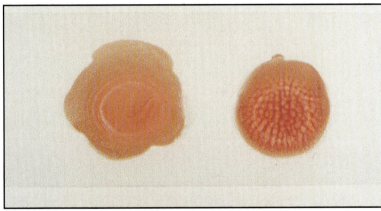

**Figure 8-25.** Rh blood type is determined by agglutination. Rh-negative blood is on the left, Rh-positive is on the right.

**Figure 8-26.** *Treponema pallidum* antigens coated on to RBCs are the basis for this hemagglutination test. Since the antigens do not naturally cause agglutination, this test is an example of an indirect agglutination. The top row consists of serially diluted standards to act as positive (reactive) controls. A positive result is evidenced by a smooth mat of cells in the well (as in well A1). A negative result is a button of cells (as in well A12). Patient samples are in rows B and C. Patients B1, B3, B4, and B5 test positive for syphilis antibodies; patient B2 is negative. Samples in row C correspond to samples in row B and are negative (nonreactive) controls. The patient's serum is mixed with unsensitized RBCs to assure that agglutination is actually due to reaction with *Treponema* antibodies.

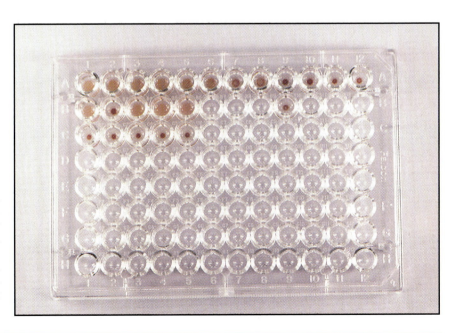

# ENZYME-LINKED IMMUNOSORBENT ASSAY (ELISA)

**Purpose**    Depending on how it is run, the ELISA may be used to detect the presence and amount of either antigen or antibody in a sample.

**Medical Applications**    The indirect ELISA is used for screening patients for the presence of HIV antibodies, rubella virus antibodies and others. The direct ELISA may be used to detect hormones (such as HCG in some pregnancy tests), drugs, and viral and bacterial antigens.

**Principle**

There are several types of ELISAs, but they basically fall into two groups: those that detect antigen in a sample, and those that detect antibody in a sample (see Fig. 8-27). All rely on a second antibody with an attached (*conjugated*) enzyme as an indicator of antigen-antibody reaction.

The *direct ELISA* detects the presence of antigen in a sample. A microtiter well is coated with *homologous antibody* specific to the antigen being looked for. The sample being assayed is added to the well. If the antigen is present, it will react with the antibody coating the well; if none is present, no reaction occurs. In this form of ELISA, the enzyme-linked antibody is also specific for the antigen being tested for. When it is added to the well, it binds to the antigen, if present. After allowing time for the enzyme-linked antibody

to react with the antigen, the well is washed out to remove any unbound enzyme-linked antibody (which would produce a false positive when the substrate is added). Upon addition of substrate, a color change is evidence of the enzyme catalyzing conversion of substrate to its product and is an indication of the sample having the antigen.

The *indirect ELISA* detects the presence of antibody in a sample. In this style of test, the microtiter well is lined with antigen specific to the antibody being looked for. The sample being assayed is added to the well and, if the antibody is present, it will react with the antigen coating the well; if none is present, no reaction occurs. In this case, the enzyme-linked antibody is an *antihuman immunoglobin antibody*—its antigen is actually an antibody! When it is added to the well, it binds to the antibody in the sample, if present. As with the direct ELISA, unbound enzyme-linked antibody must be washed away to prevent a false positive result. Enzyme substrate is added and a color change indicates a positive test.

Figures 8-28 and 8-29 illustrate a form of ELISA which is quantitative, *i.e.* it is used not only to determine the presence of antibody in a sample, but also the amount. Figure 8-30 illustrates another practical use of the ELISA technique—a home pregnancy test—that screens for the presence of *human chorionic gonadotropin* found in pregnant women.

## Direct ELISA Method

This ELISA detects the presecnce of *antigen* in a sample. Antibody (antiserum) specific for that antigen is coated (adsorbed) onto the wall of a microtiter well.

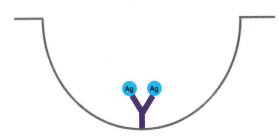

The sample is added. If the *antigen* is present, it will bind to the antibody coating the well.

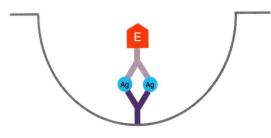

A second antibody with an attached enzyme specific for the same antigen is added. Unbound enzyme-linked antibody is washed away.

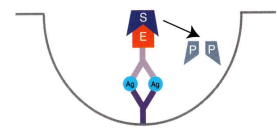

Substrate for the enzyme is added. Conversion of substrate to product is evidenced by a color change. A color change means the sample has the antigen; no color change is a negative result.

## Indirect ELISA Method

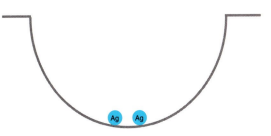

This ELISA detects the presecnce of *antibody* in a sample. The antigen specific for that antibody is coated (adsorbed) onto the wall of a microtiter well.

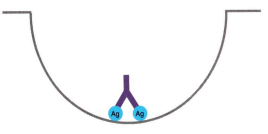

The sample is added. If the *antibody* is present, it will bind to the antigen coating the well.

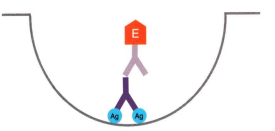

An antihuman immunoglobin antibody with an attached enzyme is added. Unbound enzyme-linked antibody is washed away.

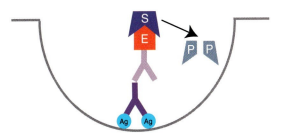

Substrate for the enzyme is added. Conversion of substrate to product is evidenced by a color change. A color change means the sample has the antibody; no color change is a negative result.

**Figure 8-27.** Direct and indirect ELISAs.

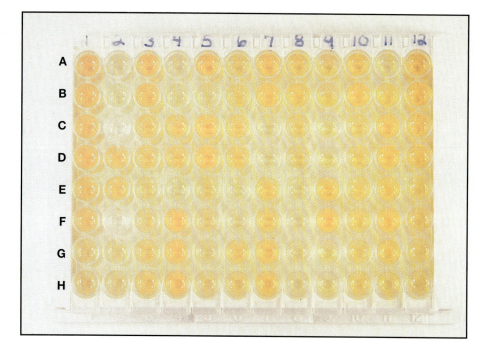

**Figure 8-28.** One example of an ELISA used to determine antibody titer. In this ELISA, a dark yellow color indicates a negative reaction. The lighter the color, the higher the antibody titer. Color is read by a photometer (see Fig. 8-29) and results are fed into a computer. A variety of controls is also used. Serially diluted antibody samples of known concentration are in column 1, rows A through H and column 2, rows A through C. Absorbance values from these are used to develop a standard curve correlating antibody titer with absorbance. Patient samples are in the other wells. Each patient's antibody titer may be determined by comparison with the standard curve.

**Figure 8-29.** Set-up for the ELISA illustrated in Figure 8-28. The photometer is in the foreground. The microtiter plate is visible on the photometer at the right. The computer is visible in the background.

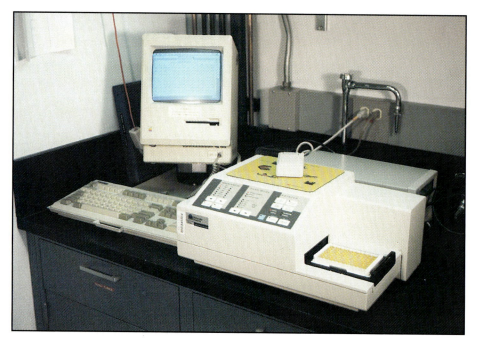

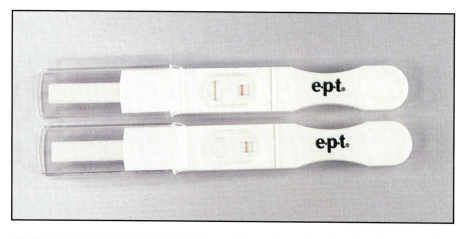

**Figure 8-30.** Another style of an ELISA is used in home pregnancy tests. Human chorionic gonadotropin (HCG) is an antigen present only in pregnant women. A sample of urine is applied to the wick on the left of the test system. Two pink lines indicate a reaction of HCG with antibodies in the test system. A single line indicates absence of HCG in the urine and is interpreted as a negative result.

# Viral, Protozoan and Fungal Microbiology

## T4 VIRUS

Viruses are not cellular organisms, having no cell membrane or metabolism of their own. Rather, they are infectious particles minimally consisting of an outer protein coat covering a nucleic acid (either RNA or DNA, not both—another significant difference between viruses and cellular life) as shown in Figure 9-1. In order to replicate (see Fig. 9-2), the virus must infect a living cell which provides the raw materials and organelles necessary for viral replication. Thus, viruses are referred to as "obligate *intracellular* parasites". Viruses that have bacterial hosts are referred to a *bacteriophages* or *phages*.

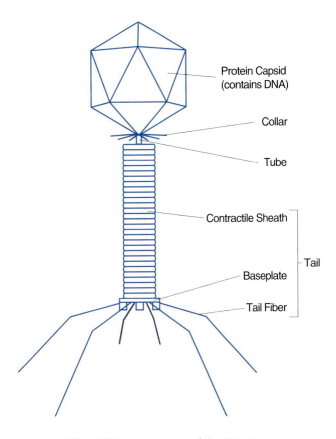

Protein Capsid
(contains DNA)

Collar

Tube

Contractile Sheath

Tail

Baseplate

Tail Fiber

*Figure 9-1.* A diagram of the T4 virus.

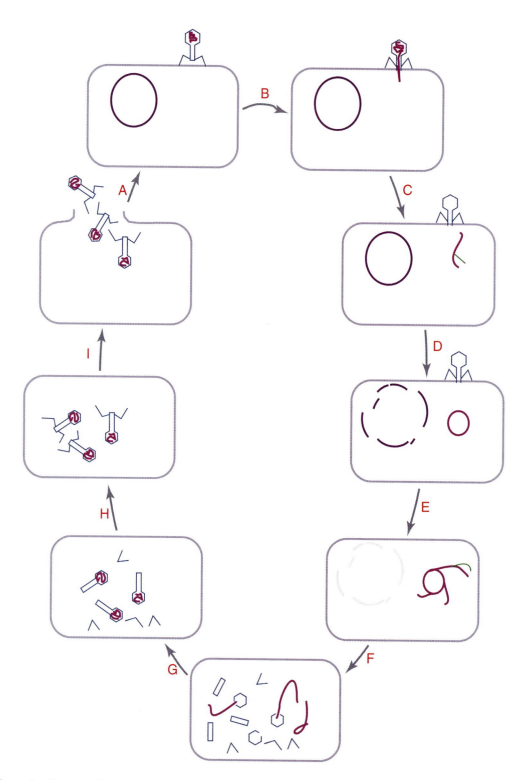

**Figure 9-2.** A schematic diagram of the T4 replicative cycle. *A.* The virus particle (shown in blue) attaches to the host cell (shown in gray). Each virus will infect only a specific host and this specificity is based on the ability of the virus to make attachment with viral receptors on the host. *B.* The virus particle acts like a syringe and injects its DNA (shown in violet) into the host cell. *C.* Viral DNA is not transcribed all at once. Rather, the genes necessary for the early events in replication are transcribed first, with other genes being transcribed as the appropriate time arises. In this diagram, *early* and *middle genes* are being transcribed into mRNA (shown in green). *D.* The viral DNA circularizes and the host DNA is broken apart. *E.* Viral DNA is replicated by a *rolling circle* mechanism. Host DNA is further degraded into nucleotides which are used for viral DNA replication. *Late genes* are also transcribed. *F.* Capsid, tail and tail fiber proteins are made and assemble into their respective components in separate assembly lines. DNA enters the capsids. *G & H.* Assembly continues as first tails and then tail fibers attach to form a complete virus particle (*virion*). *I.* The host cell is lysed and releases the completed virus particles, each capable of infecting another host cell. The typical *burst size* for T4 is about 300 viruses. The whole process from attachment to host lysis takes less than 25 minutes!

# HUMAN IMMUNODEFICIENCY VIRUS (HIV)

HIV is the causative agent of AIDS (*Acquired Immune Deficiency Syndrome*). At first, only a single type of HIV was known, but in 1985 a second HIV was isolated. The two forms are now referred to as HIV-1 and HIV-2, respectively.

HIV (see Figs. 9-3 and 9-4) is a *retrovirus*. It has a spherical capsid enclosed by a phospholipid envelope derived from the host cell membrane. Within it is a protein core surrounding two single stranded RNA molecules, each of which is associated with a molecule of *reverse transcriptase*. The HIV genome consists of only 9 genes. Glycoprotein spikes emerging from the envelope are involved in attachment to the host cell.

HIV infects cells with CD4 membrane receptors, normally used for antigen recognition, but used by HIV for attachment. A subpopulation of T cells, the $T_4$ helper cells, are most commonly infected and die as a result. Other cells, such as dendritic cells–a class of antigen presenting cells, macrophages and monocytes may also be infected. After infection, the reverse transcriptase catalyzes the production of viral DNA (provirus) which integrates into the host chromosome and is the template for production of viral RNA and proteins. After assembly, new virions emerge from the host cell by budding and infect other cells. Latent infection, in which no new virus is made, is also a possible outcome of infection.

$T_4$ helper cells are essential to the normal operation of the immune system since they promote development of immune cells in both humoral and cell-mediated responses. Depletion of $T_4$ cells cripples the immune system and the patient becomes susceptible to infections by organisms not typically pathogenic to humans. Thus, AIDS is not a single disease, but rather a *syndrome* of diseases characteristic of patients with HIV infection.

HIV is transmitted via body fluids such as blood, breast milk, semen, and vaginal secretions. Infection can occur as a result of sexual intercourse with or blood transfusion from an infected individual. Infection may also occur across the placenta during pregnancy or via contaminated needles used for injection of intravenous drugs. It also may be transmitted to a newborn during delivery or nursing by an infected mother. Casual social contact does not appear to be a route of infection.

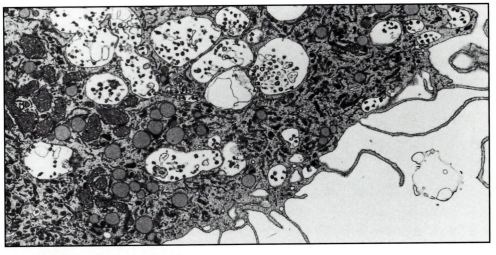

**Figure 9-3.** Electron micrograph of a macrophage from an HIV infected person (X7200). Vacuoles contain viral particles. (Photo courtesy of Dr. Rachel Schrier and Dr. Clayton Wiley.)

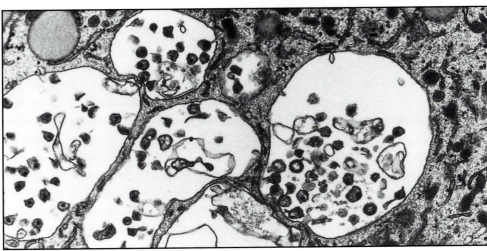

**Figure 9-4.** High power view of virus particles in Figure 9-3 (X22,500). (Photo courtesy of Dr. Rachel Schrier and Dr. Clayton Wiley.)

# PROTOZOAN SURVEY

Protozoans are unicellular eukaryotic heterotrophic microorganisms. Traditionally, they are divided into four classes: Sarcodina, Mastigophora, Sporozoa and Ciliata. Sarcodines move by sending out extensions of cytoplasm called *pseudopods*. Division is by binary fission. Ciliates owe their motility to the numerous cilia covering the cell. Reproduction is by transverse fission. Members of Mastigophora are characterized by one or more flagella and division by longitudinal fission. Sporozoans are typically nonmotile and usually have complex life cycles involving asexual reproduction in one host and sexual reproduction in another.

Figures 9-5 through 9-8 show nonpathogenic representatives of Mastigophora, Sarcodina, and Ciliata. Pathogenic species are shown in the section *Protozoans of Medical Importance*.

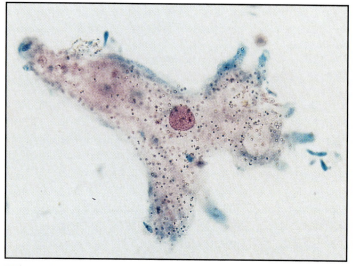

**Figure 9-5.** *Amoeba*, a sarcodine (X212). Note the numerous pseudopods.

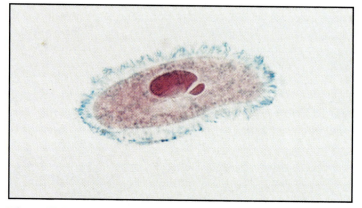

**Figure 9-6.** *Paramecium bursaria*, a ciliate (X528). Note the cilia around the edge of the cell. The macronucleus and micronucleus are also visible.

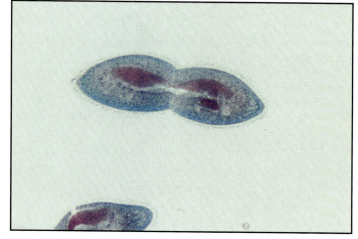

**Figure 9-7.** *Paramecium* undergoing transverse fission (X264).

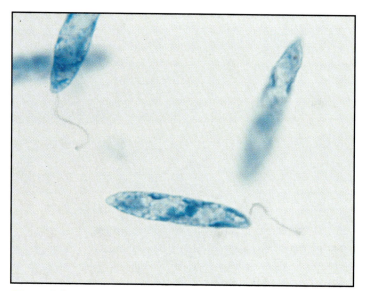

**Figure 9-8.** *Euglena*, a flagellate (X1584).

## Amoeboid Pathogen

**Entamoeba histolytica**    *Entamoeba histolytica* is the causative agent of *amoebic dysentery* (*amebiasis*), a disease most common in areas with poor sanitation. The organism exists in two forms: a vegetative *trophozoite* and a *cyst* stage (see Figs. 9-9 and 9-10).

Infection occurs when cysts are ingested by a human host, either through fecal-oral contact or more typically, contaminated food or water. Cysts (but not trophozoites) are able to withstand the acidic environment of the stomach. Upon entering the less acidic small intestine, the cysts undergo *excystation*. Mitosis produces eight small trophozoites from each cyst.

The trophozoites parasitize the mucosa and submucosa of the colon causing ulcerations. They feed on red blood cells and bacteria. The extent of damage determines whether the disease is acute, chronic, or asymptomatic. In the most severe cases, infection may extend to other organs, especially the liver, lungs or brain. Abdominal pain, diarrhea, blood and mucus in feces, nausea, vomiting, and hepatitis are among the symptoms of amebic dysentery.

Cysts develop when the fecal material becomes too solid to be favorable for the trophozoites. Initially uninucleate, mitosis produces the mature quadranucleate cyst. Cysts are shed in the feces and may infect new hosts. They may also persist in the original host resulting in an *asymptomatic carrier*—a major source of contamination and infection.

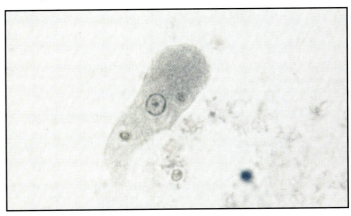

**Figure 9-9.** *Entamoeba histolytica* trophozoite (X800).

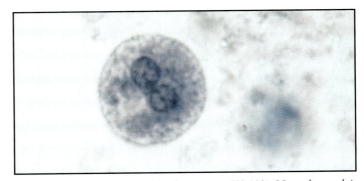

**Figure 9-10.** *Entamoeba histolytica* cyst (X2640). Note the nuclei.

## Ciliate Pathogen

**Balantidium coli**    *Balantidium coli* (see Fig. 9-11) is the causative agent of *balantidiasis* and exists in two forms: a vegetative trophozoite and a cyst.

The trophozoite is highly motile due to the cilia and has a macro- and a micronucleus. Cysts in sewage-contaminated water are the infective form. Trophozoites may cause ulcerations of the colon mucosa, but not to the extent produced by *Entamoeba histolytica*. Symptoms of acute infection are bloody and mucoid feces. Diarrhea alternating with constipation may occur in chronic infections. Most infections are probably asymptomatic.

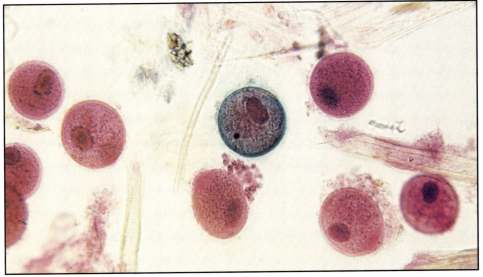

**Figure 9-11.** *Balantidium coli* (X264).

## Flagellate Pathogens

**Giardia lamblia** *Giardiasis* is caused *Giardia lamblia*, a flagellate protozoan. It exists in the duodenum as a heart-shaped vegetative trophozoite (see Fig. 9-12) with four pairs of flagella and a sucking disc that allows it to resist gut peristalsis. Multinucleate cysts (Fig. 9-13) are formed when the organism enters the colon. Cysts are shed in the feces and may produce infection of a new host upon ingestion. Transmission typically involves fecally contaminated water or food, but direct fecal-oral contact transmission is also possible.

The organism attaches to epithelial cells, but does not penetrate to deeper tissues. Most infections are asymptomatic. Chronic diarrhea, dehydration, abdominal pain and other symptoms may occur if the infection produces a large enough population to involve a significant surface area of the small intestine.

**Trichomonas vaginalis** *Trichomonas vaginalis* is the causative agent of *trichomoniasis (vulvovaginitis)* in humans (see Fig. 9-14). It has four anterior flagella and an undulating membrane.

Trichomoniasis may affect both sexes, but is more common in females. *T. vaginalis* causes inflammation of genitourinary mucosal surfaces—typically the vagina, vulva and cervix in females and the urethra, prostate and seminal vesicles in males. Most infections are asymptomatic or mild. There may be some erosion of surface tissues and a discharge associated with infection. The degree of infection is affected by host factors, especially the bacterial flora present and the pH of the mucosal surfaces. Transmission typically is by sexual intercourse.

The morphologically similar nonpathogenic *Trichomonas tenax* and *T. hominis* are residents of the oral cavity and intestines, respectively.

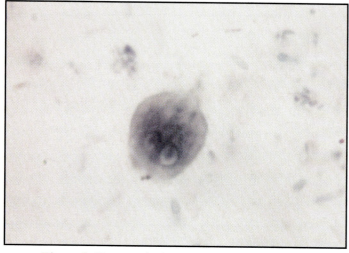

**Figure 9-12.** *Giardia lamblia* trophozoite (X2640).

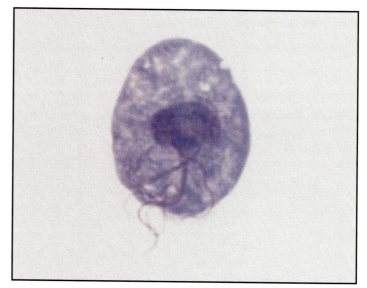

**Figure 9-14.** *Trichomonas vaginalis* (X2534).

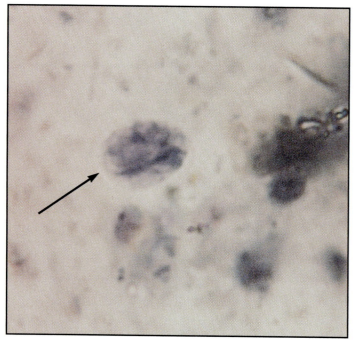

**Figure 9-13.** *Giardia lamblia* cyst (X2960).

**Leishmania donovani** *Leishmania donovani* actually represents a number of geographically separate species and subspecies that are difficult to distinguish morphologically. All produce *visceral leishmaniasis* or *kala-azar*, a disease found in tropical and subtropical regions. The organism exists as a nonflagellated *amastigote* in the mammalian host (humans, dogs, and rodents) and as an infective, motile *promastigote* in the sandfly vector (see Fig. 9-15). They are introduced into the mammalian host by sandfly bites. Distribution of the disease is associated with distribution of the appropriate sandfly vector.

Upon introduction into the host by the sandfly, the organism is phagocytized by macrophages and converts to the amastigote stage. Mitotic divisions result in filling of the macrophage, which bursts and releases the parasites. Phagocytosis by other macrophages follows and the process repeats. In this way, the organism spreads through much of the reticuloendothelial system, including lymph nodes, liver, spleen and bone marrow. Kala-azar is a progressive disease and is fatal if untreated.

Amastigotes in an infected host may be ingested by a sandfly during a blood meal. Once inside the sandfly, they develop into the promastigote form and multiply. They eventually occupy the fly's buccal cavitiy where they can be transmitted to a new mammalian host during a subsequent blood meal. Transmission requires the vector and does not occur by direct contact.

The related *Leishmania tropica* and *L. mexicana* cause *Oriental sore*, a cutaneous infection. *L. braziliensis* causes *New World cutaneous leishmaniasis*, an infection of skin and oral, nasal and pharyngeal mucous membranes. In all cases, infection involves macrophages in the affected region. Unlike *L. donovani*, all three may be transmitted by direct contact or by sandfly bites. Leishmaniasis was a common infection among troops in the Gulf War.

**Trypanosoma brucei** *Trypanosoma brucei* (see Fig. 9-16) is a species of flagellated protozoans divided into two subspecies: *T. brucei gambiense* and *T. brucei rhodesiense*. Both produce *African trypanosomiasis*, also known as *African sleeping sickness*. They are very similar morphologically, but differ in geographic range.

Trypanosomes have a complex life cycle. One stage of the life cycle, the *epimastigote*, multiplies in an intermediate host, the tsetse fly (genus *Glossina*). The infective *trypomastigote* stage is then transmitted to the human host through tsetse fly bites. Once introduced, trypomastigotes multiply and produce a chancre at the site of the bite. They enter the lymphatic system and spread through the blood, ultimately to the heart and brain. Immune response to the pathogen is hampered by the trypanosome's ability to change surface antigens faster than the immune system can produce appropriate antibodies. This antigenic variation also makes development of a vaccine unlikely.

Progressive symptoms include headache, fever and anemia, followed by symptoms characteristic of the infected sites. The sleeping sickness symptoms–sleepiness, emaciation, and unconsciousness–begin when the central nervous system becomes infected. The disease may last for years, but mortality rate is high. Death results from heart failure, meningitis, or severe debility of some other organ(s).

The infective cycle is complete when an infected individual (humans, cattle, and some wild animals are reservoirs) is bitten by a tsetse fly which ingests the organism during its blood meal. It becomes infective for its lifespan.

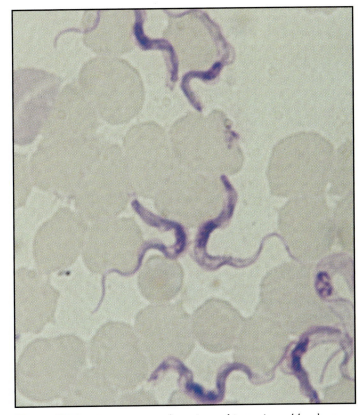

**Figure 9-16.** *Trypanosoma brucei gambiense* in a blood smear (X2640).

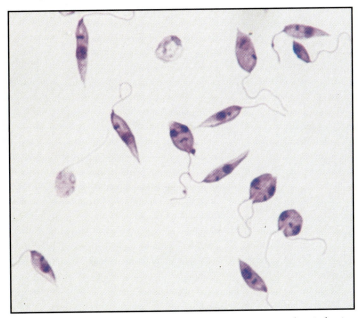

**Figure 9-15.** *Leishmania donovani* promastigote, the infective stage, obtained from a culture (X1320).

**Trypanosoma cruzi** *Trypanosoma cruzi* (see Fig. 9-17) causes *American trypanosomiasis (Chagas' disease)*. Cone-nosed ("kissing") bugs are the insect vector. They transmit the infective trypanosome during a blood meal through their feces. Scratching introduces the organism into the bite wound or conjunctiva. A local lesion (*chagoma*) forms at the entry site and is accompanied by fever. Spreading occurs via lymphatics (producing *lymphadenitis*) and trypomastigotes may be found in the blood within a couple of weeks. Trypanosomes then become localized in reticuloendothelial cells of the spleen, liver and bone marrow where they multiply intracellularly. Infected individuals may infect the cone-nosed bugs during a subsequent blood meal.

American trypanosomiasis occurs in South and Central America. It may be fatal, mild, or asymptomatic in adults. It is especially severe in children who often introduce the trypanosome through the conjunctiva, leading to edema of the eyelids and face of the affected side. The disease may spread to the central nervous system or to the heart, causing severe myocarditis.

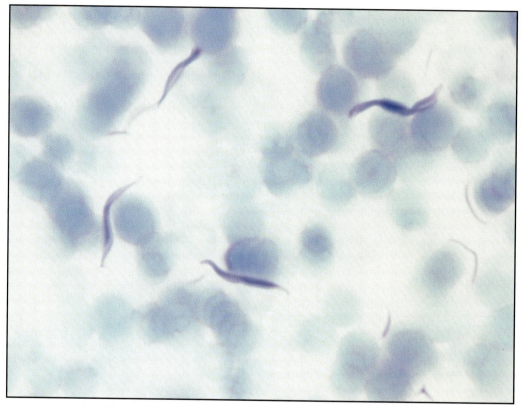

*Figure 9-17.* *Trypanosoma cruzi* in a blood smear (X1742).

*A Photographic Atlas for the Microbiology Laboratory*

**Plasmodium spp.** Plasmodia are sporozoan parasites with a complex life cycle, part of which is in various vertebrate tissues, while the other part involves an insect. In humans, the tissues are the liver and red blood cells, while the insect vector is the female *Anopheles* mosquito. The life cycle is shown in Figure 9-18.

There are four species of *Plasmodium* that cause malaria in humans. These are *P. vivax* (benign tertian malaria), *P. malariae* (quartan malaria), *P. falciparum* (malignant tertian malaria), and *P. ovale* (ovale malaria). The life cycles are similar for each species as is the progress of the disease, so *P. falciparum* will be discussed as an example, with unique aspects compared to the others.

The *sporozoite* stage of the pathogen is introduced into a human host during a bite from an infected female *Anopheles* mosquito. Sporozoites then infect liver cells and produce the asexual *merozoite* stage. Merozoites are released from lysed liver cells, enter the blood, and infect erythrocytes. (Reinfection of the liver occurs at this stage in all but *P. falciparum* infections.) Once in RBCs, merozoites enter a cyclic pattern of reproduction in which more merozoites are released from the red cells synchronously every 48 hours (hence *tertian*–every third day–malaria). These events are tied to the symptoms of malaria. A chill, nausea, vomiting and headache are symptoms that correspond to rupture of the erythrocytes. A spiking fever ensues and is followed by a period of sweating, after which the exhausted patient falls asleep. It is during this latter phase that the parasites reinfect the red cells, and the cycle repeats.

The sexual phase of the life cycle begins when certain merozoites enter erythrocytes and differentiate into male or female *gametocytes*. The sexual phase of the life cycle continues when ingested by a female *Anopheles* mosquito during a blood meal. Fertilization occurs and the zygote eventually develops into a cyst within the mosquito's gut wall. After many divisions, the cyst releases sporozoites, some of which enter the mosquito's salivary glands ready to be transmitted back to the human host.

Most malarial infections eventually are cleared, but not before the patient has developed anemia and has suffered permanent damage to the spleen and liver. The most severe infections involve *P. falciparum*. Erythrocytes infected by *P. falciparum* develop abnormal projections that cause them to adhere to the lining of small blood vessels. This can lead to obstruction of the vessels, thrombosis, or local ischemia which account for many of the fatal complications of this type of malaria–including liver, kidney and brain damage.

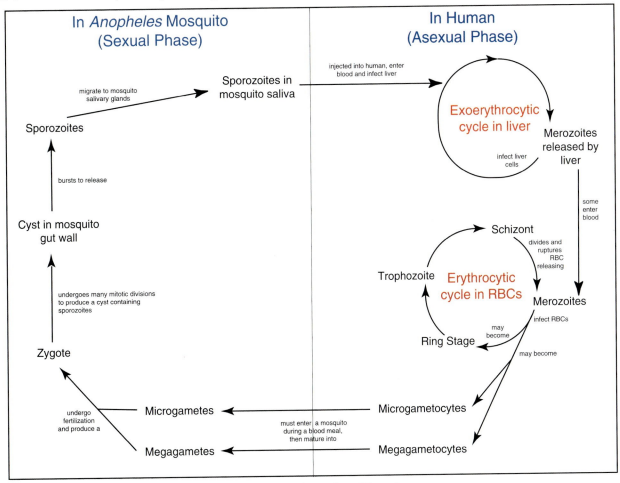

**Figure 9-18.** *Plasmodium* life cycle.

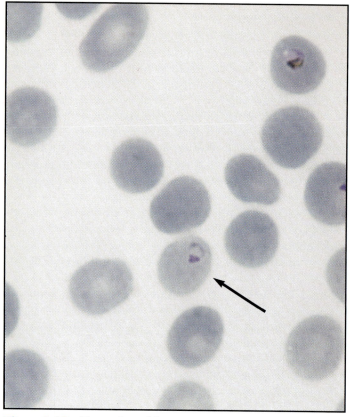

**Figure 9-19.** *Plasmodium falciparum* ring stage in a red blood cell (X2640).

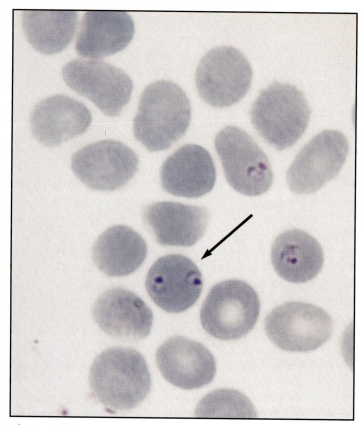

**Figure 9-20.** *Plasmodium falciparum* double infection of a red blood cell (X2640).

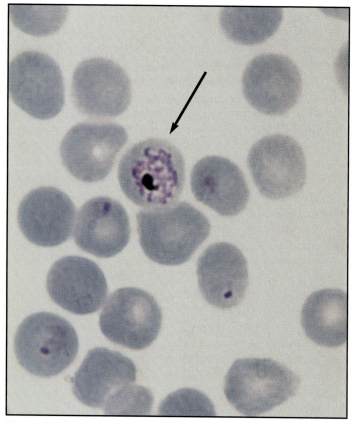

**Figure 9-21.** *Plasmodium falciparum* developing schizont in a red blood cell (X2640).

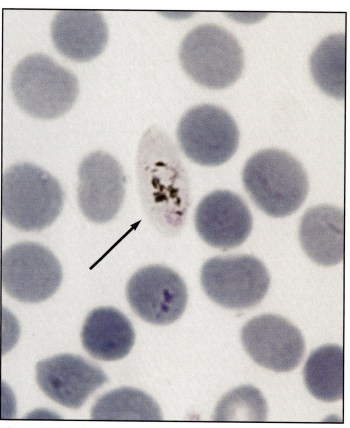

**Figure 9-22.** *Plasmodium falciparum* gametocyte in a blood smear (X2640).

*A Photographic Atlas for the Microbiology Laboratory*

**Toxoplasma gondii** Like other sporozoans, *Toxoplasma gondii* (Figure 9-23) has sexual and asexual phases in its life cycle. The sexual phase occurs in the lining of cat intestines where *oocysts* are produced and shed in the feces. Each oocyst undergoes division and contains 8 sporozoites. If ingested by another cat, the sexual cycle may be repeated as the sporozoites produce gametocytes which in turn produce gametes. If ingested by another animal host (including humans) the oocyst germinates in the duodenum and releases the sporozoites. Sporozoites enter the blood and infect other tissues where they become trophozoites, which continue to divide and spread the infection to lymph nodes and other parts of the reticuloendothelial system. Trophozoites ingested by a cat from eating an infected animal develop into gametocytes in the cat's intestines. Gametes are formed, fertilization produces an oocyst, and the life cycle is completed.

Infection via ingestion of the oocyst typically is not serious. The patient may notice fatigue or muscle aches. The more serious form of the disease involves infection of a fetus across the placenta from an infected mother. This type of infection may result in stillbirth, or liver damage and brain damage. AIDS patients may also suffer fatal complications from infection.

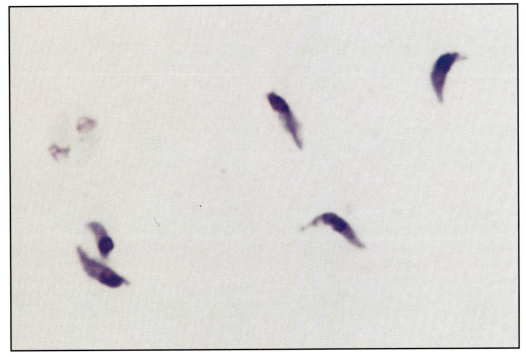

**Figure 9-23.** *Toxoplasma gondii* (X3168).

# SURVEY OF FUNGI

Fungi are eukaryotic absorptive heterotrophs. (They produce exoenzymes that digest nutrients in the environment, then absorb the products rather than ingest their food like animals.) Most are *saprophytes* that decompose dead organic matter, but some are *parasites* of plants, animals, or humans.

The fungi are unique enough to be designated to their own kingdom. Fungi are somewhat informally divided into unicellular *yeasts* and filamentous *molds* based on their overall appearance. Filamentous fungi that produce fleshy reproductive structures—mushrooms, puffballs, and shelf fungi—are referred to as *macrofungi* (even though the majority of the fungus is filamentous and hidden underground or within a decaying tree).

Fungal filaments are called *hyphae*. Collectively, the hyphae of a fungus form a *mycelium*. Gametes are produced by *gametangia*. Spores are produced by a variety of *sporangia*. Typically, the only diploid cell in the fungal life cycle is the zygote, which then undergoes meiosis to produce haploid spores characteristic of the fungal group. Asexual spores may also be produced during the life cycle of many fungi.

More formal taxonomic categories based primarily on the pattern of sexual spore production are also used. Zygomycetes are terrestrial and produce nonmotile sporangiospores and zygospores. Ascomycetes produce a sac (an *ascus*) in which the zygote undergoes meiosis to produce haploid *ascospores*. Basidiomycetes that undergo sexual reproduction produce a *basidium* which undergoes meiosis to produce four *basidiospores* which are attached to its surface. Deuteromycetes are an unnatural assemblage of fungi in which sexual stages are either unknown or are not used in classification. The majority of deuteromycetes resemble ascomycetes.

Following is a survey of fungi likely to be encountered in an introductory microbiology class.

## Yeasts

**Saccharomyces cerevisiae**    *Saccharomyces cerevisiae* is an ascomycete used in production of bread, wine and beer. It does not form a mycelium, but rather produces a colony not unlike those of bacteria (see Fig. 9-24). Asexual reproduction occurs by budding. Meiosis produces four ascospores within the vegetative cell which acts as the ascus (see Fig. 9-26). Ascospores may fuse to form another generation of diploid vegetative cells or they may be released to produce a population of haploid cells that are indistinguishable from diploid cells. Haploid cells of opposite mating types may also combine to create a diploid cell.

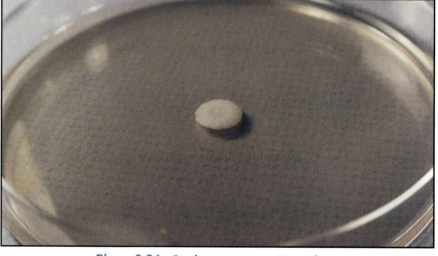

**Figure 9-24.** *Saccharomyces cerevisiae* colony.

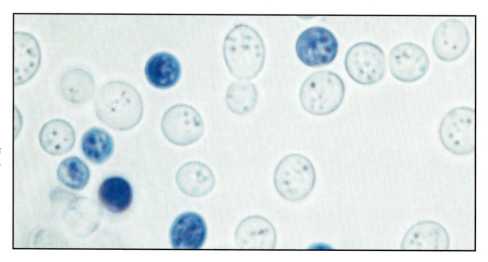

**Figure 9-25.** Wet mount of *Saccharomyces cerevisiae* vegetative cells (X2640). Note budding cell in the center of field.

**Figure 9-26.** *Saccharomyces cerevisiae* ascospores (X1320).

**Candida albicans**  *Candida albicans* (see Fig. 9-27) is part of the normal respiratory, gastrointestinal and female urogenital tract floras. Under the proper circumstances, it may flourish and produce pathological conditions, such as *thrush* in the oral cavity, *vulvovaginitis* of the female genitals, and *cutaneous candidiasis* of the skin. *Systemic candidiasis* may follow infection of the lungs, bronchi or kidneys. Entry into the blood may result in *endocarditis*. Individuals most susceptible to *Candida* infections are diabetics, those with immunodeficiency (*e.g.* AIDS), catheterized patients, and individuals taking antimicrobial medications.

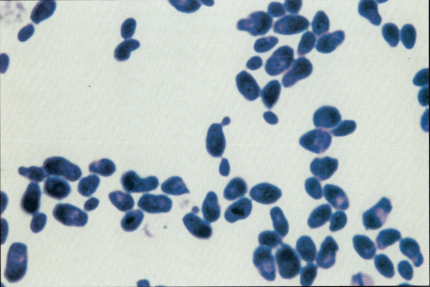

**Figure 9-27.** *Candida albicans* vegetative cells (X2640).

## Molds

**Rhizopus stolonifer**  *Rhizopus stolonifer* (see Figs. 9-28 and 9-29) is the common bread mold and is a member of the Zygomycetes. Its hyphae are nonseptate (*coenocytic*), and cytoplasmic streaming within them is common. Surface hyphae (*stolons*) are anchored by *rhizoids* where the hyphae contact the substrate.

The life cycle is illustrated in Figure 9-30. Hyphae are haploid. Asexual spores are produced by sporangia (see Fig. 9-31) at the ends of elevated *sporangiophores*. These spores develop into hyphae identical to those that produced them. On occasion, sexual reproduction occurs when hyphae of different mating types (+ and - strains) make contact. Initially, *progametangia* (see Fig. 9-32) extend from each hypha. Upon contact, a septum separates the end of each progametangium into a gamete (see Fig. 9-33). The walls between the two gametangia dissolve and a thick-walled *zygosporangium* develops (see Figs. 9-34 and 9-35). Fusion of nuclei occurs within the zygosporangium and produces one or more diploid nuclei, or *zygotes*. The zygosporangium then germinates and a sporangium similar to the asexual sporangia develops. Meiosis produces haploid spores which develop into new hyphae, and the life cycle is completed.

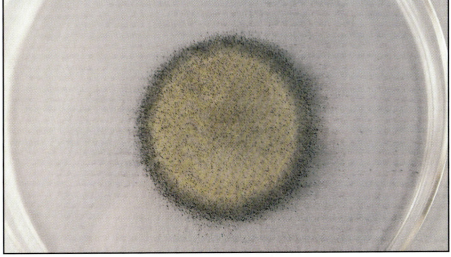

**Figure 9-28.** *Rhizopus stolonifer* colony. The black sporangia are visible.

**Figure 9-29.** *Rhizopus stolonifer* from underneath before producing sporangia. Its filamentous nature is apparent.

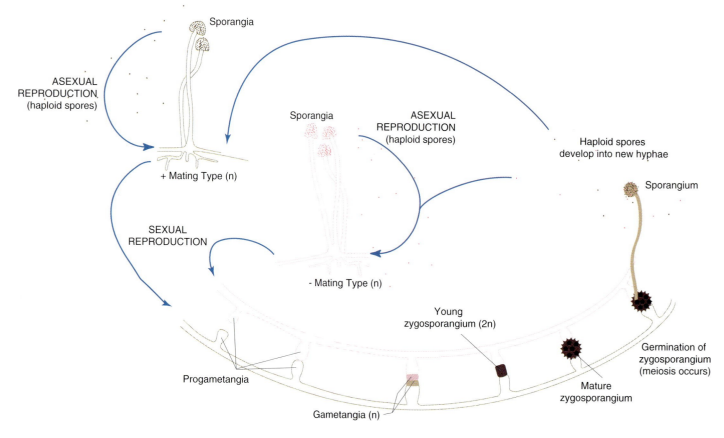

Sporangia

ASEXUAL
REPRODUCTION
(haploid spores)

+ Mating Type (n)

Sporangia

ASEXUAL
REPRODUCTION
(haploid spores)

Haploid spores
develop into new hyphae

Sporangium

SEXUAL
REPRODUCTION

- Mating Type (n)

Young
zygosporangium (2n)

Germination of
zygosporangium
(meiosis occurs)

Progametangia

Mature
zygosporangium

Gametangia (n)

**Figure 9-30.** *Rhizopus* life cycle.

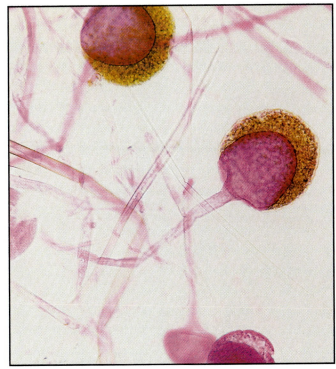

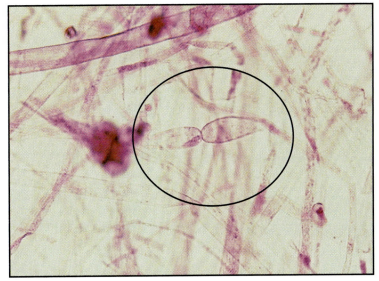

**Figure 9-32.** *Rhizopus* progametangia from different hyphae shown in the center of the field (X264).

**Figure 9-31.** *Rhizopus* sporangiophore (X264). The gold sporangium contains asexual spores.

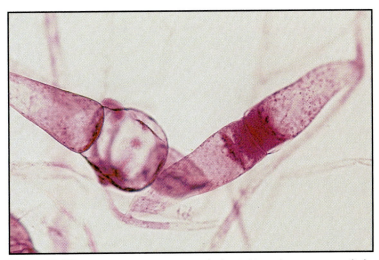

**Figure 9-33.** *Rhizopus* gametangia (dark pink) and suspensors (light pink) in the center of the field (X264).

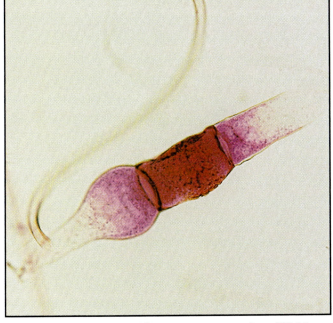

**Figure 9-34.** Young *Rhizopus* zygosporangium (X264).

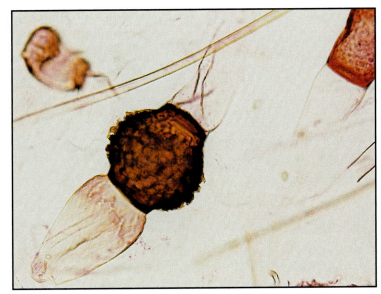

**Figure 9-35.** Mature *Rhizopus* zygosporangium (zygospore) (X264).

**Penicillium** Members of the genus *Penicillium* (see Figs. 9-36 and 9-37) are classified by some as deuteromycetes because their sexual reproductive structures are not known or are not used in classification. However, it is clear that they are actually ascomycetes. Sexual reproduction results in the formation of ascospores within an ascus (see Figs. 9-38 and 9-39). Asexual reproduction occurs via *conidia*, small spores not enclosed within a sporangium (see Fig. 9-40).

*Penicillium* is best known for its production of the antibiotic penicillin. Other species of *Penicillium* are of commercial importance for fermentations used in cheese production. Examples include *P. roquefortii* (Roquefort cheese) and *P. camembertii* (Camembert and Brie cheeses).

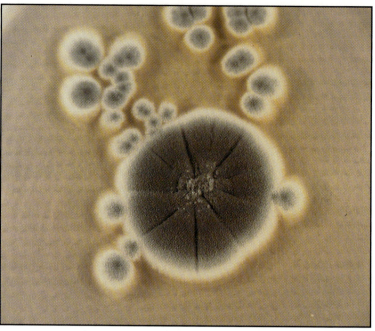

**Figure 9-36.** *Penicillium notatum* colony.

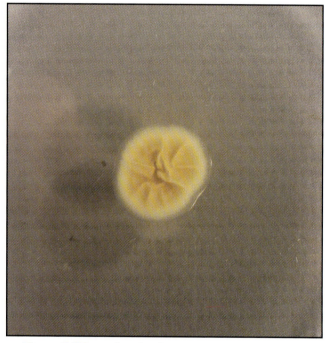

**Figure 9-37.** Underside of *Penicillium notatum* colony.

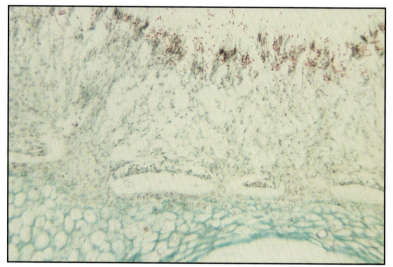

**Figure 9-38.** Section through *Penicillium* (X132). Red ascospores are visible at the upper surface.

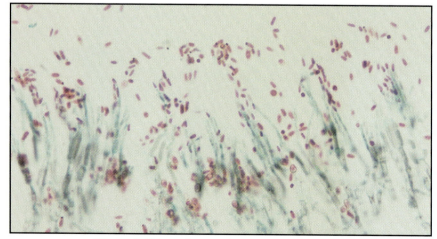

**Figure 9-39.** Red *Penicillium* ascospores (X528).

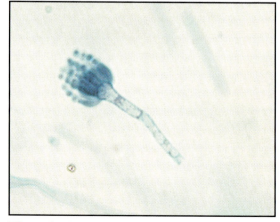

**Figure 9-40.** *Penicillium* conidiophore with chains of asexual spores (conidia) at the end (X792).

*A Photographic Atlas for the Microbiology Laboratory*

**Aspergillus** Species of *Aspergillus*, like *Penicillium*, are considered deuteromycetes by some and ascomycetes by others (see Figs. 9-41 and 9-42). *A. fumigatus* and other species are opportunistic pathogens that cause *aspergillosis*, an umbrella term covering many diseases. One form of pulmonary aspergillosis (referred to as *fungus ball*) involves colonization of the bronchial tree or tissues damaged by tuberculosis. *Allergic aspergillosis* may occur in individuals who are in frequent contact with the spores and become sensitized to them. Subsequent contact produces symptoms similar to asthma. *Invasive aspergillosis* is the most severe form. It results in necrotizing pneumonia and may spread to other organs.

Some species of *Aspergillus* are of commercial importance. Fermentation of soybeans by *A. oryzae* produces soy paste. Soy sauce is produced by fermenting soybeans with a mixture of *A. oryzae* and *A. soyae*. *Aspergillus* is also used in commercial production of citric acid.

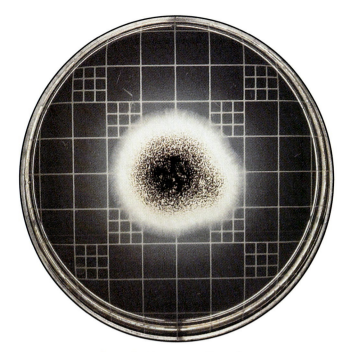

**Figure 9-41.** *Aspergillus* colony.

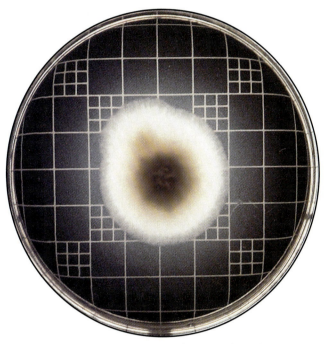

**Figure 9-42.** Underside of *Aspergillus* colony.

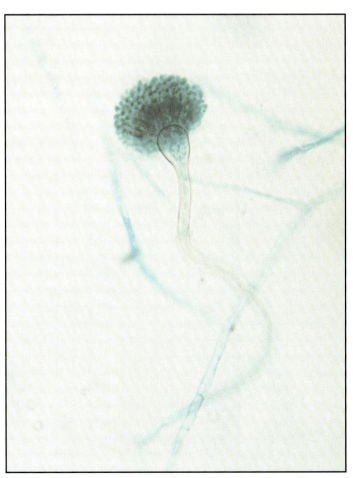

**Figure 9-43.** *Aspergillus* conidiophore with chains of asexual spores (conidia) at the end (X634).

**Pneumocystis carinii**   *Pneumocystis carinii* is an opportunistic pathogen. It produces AIDS-related pneumocystis pneumonia (see Fig. 9-44) and was a leading cause of death in AIDS patients prior to development of prophylactic medications. Before the AIDS epidemic, *P. carinii* was only known to produce interstitial plasma cell pneumonitis in malnourished infants and immunosuppressed individuals.

Comparison of ribosomal RNA has led to the conclusion that *Pneumocystis carinii*, long classified as a protozoan, is more closely related to the fungi. Its protozoan "roots" are still evident, however, in the terminology associated with it. *P. carinii* exists as a trophozoite and a multinucleate cyst (see Fig. 9-45).

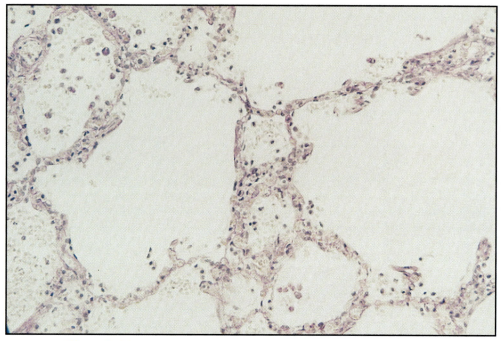

**Figure 9-44.** Section of lung infected with *Pneumocystis carinii* (X264).

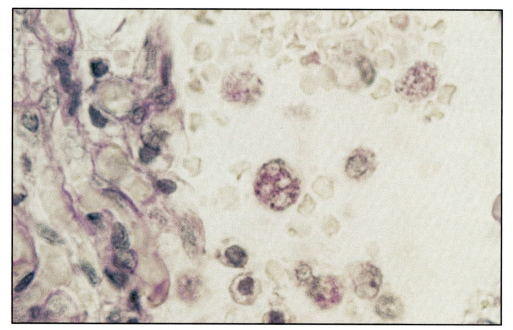

**Figure 9-45.** Section of *Pneumocystis carinii* (X1264).

# Glycolysis

The glycolytic pathway (shown in black with red enzymes in Figure A-1) is used in energy metabolism. Each glucose oxidized in glycolysis yields two pyruvic acids, 2 NADH+H⁺, and a net of 2 ATPs. The NADH+H⁺ may be oxidized in an electron transport chain or a fermentation pathway, depending on the organism and the environmental conditions. The former yields ATP, the latter does not.

Though its intermediates are carbohydrates, many are entry points for amino acid, lipid, and nucleotide catabolism (shown in purple). Many of the intermediates are also a source of carbon skeletons for the synthesis of these other biochemicals (shown in green). *Details have been omitted from these other pathways. Single arrows may represent several reactions and other carbon compounds not illustrated may be required to complete a particular reaction.*

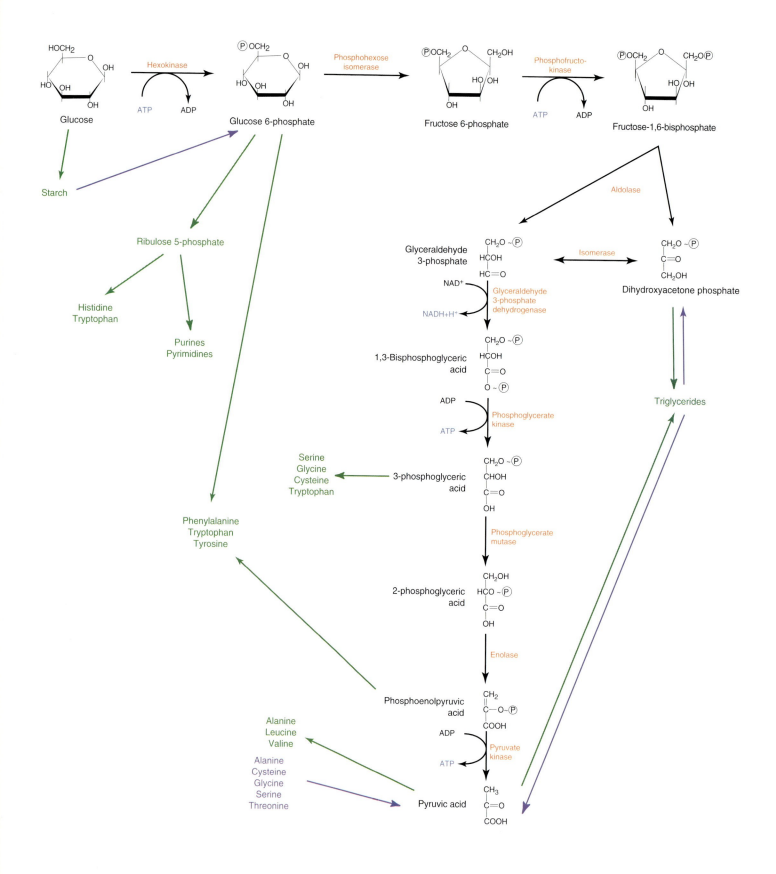

**Figure A-1.** Glycolysis and associated pathways.

# Fermentation Pathways

Figure B-1 illustrates some major fermentation pathways exhibited by microbes (though no single organism is capable of all of them—see the Table below). Pyruvic acid (shown in the blue box) is typically the starting point for each. End products of fermentation are shown in red. Fermentation allows a cell living under anaerobic conditions to oxidize reduced coenzymes (such as NADH+H$^+$ and shown in blue) generated during glycolysis or other pathways. Some bacteria (aerotolerant anaerobes) rely solely on fermentation and do not use oxygen even if it is available.

Notice that fermentation end products typically fall into three categories: acid, gas, or an organic solvent (an alcohol or a ketone). The specific fermentation performed is the result of the enzymes present in a species and is often used as a basis of classification.

| FERMENTATION | MAJOR END PRODUCTS | REPRESENTATIVE ORGANISMS |
|---|---|---|
| Alcoholic Fermentation | Ethanol and $CO_2$ | *Saccharomyces cerevisiae* |
| Homofermentation | Lactic acid | *Streptococcus* and some *Lactobacillus* |
| Heterofermentation | Lactic acid, ethanol, and acetate | *Streptococcus, Leuconostoc* and *Lactobacillus* |
| Mixed Acid Fermentation | Acetic acid, formic acid, succinic acid, $CO_2$, $H_2$ and ethanol | *Escherichia, Salmonella, Klebsiella* and *Shigella* |
| 2,3-Butanediol Fermentation | 2,3-Butanediol | *Enterobacter, Serratia* and *Erwinia* |
| Butyric Acid/Butanol Fermentation | Butanol, butyric acid, acetone and isopropanol | *Clostridium, Butyrivibrio* and some *Bacillus* |
| Propionic Acid Fermentation | Propionic acid, acetic acid and $CO_2$ | *Propionibacterium, Veillonella* and some *Clostridium* |

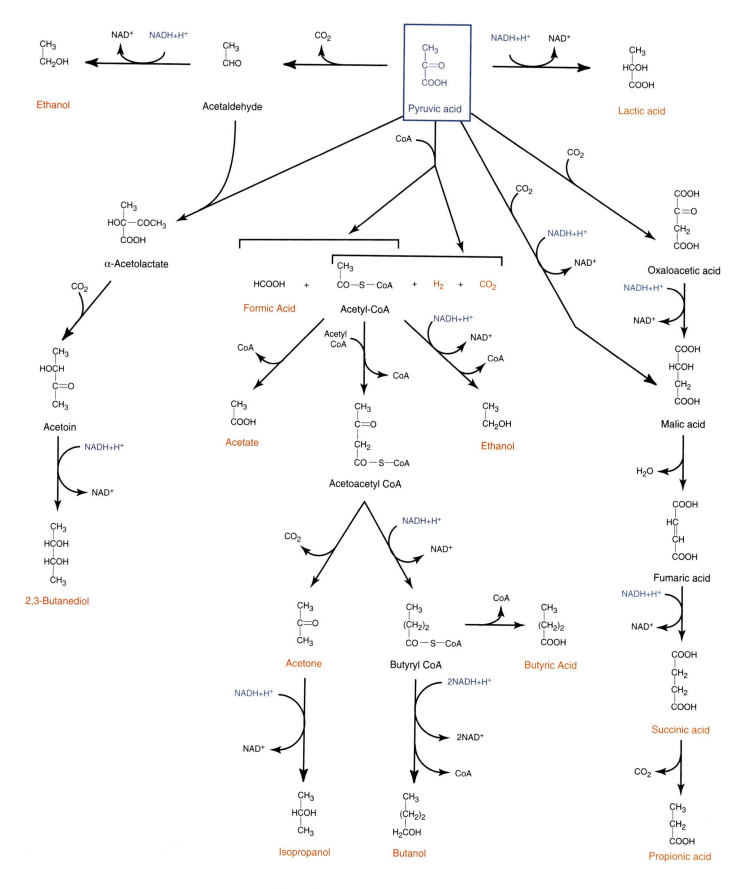

**Figure B-1.** Fermentation pathways.

# Krebs Cycle

The Krebs cycle (see Fig. C-1) is a major metabolic pathway used in energy production. Pyruvic acid produced in glycolysis or other pathways is first converted to acetyl-coenzyme A during the *entry step*. Acetyl-CoA enters the Krebs cycle through a condensation reaction with oxaloacetic acid. Products for each pyruvic acid that enters the cycle via the entry step are: 3 $CO_2$, 4 NADH+$H^+$, 1 $FADH_2$, and 1 GTP. The energy released from oxidation of reduced coenzymes (NADH+$H^+$ and $FADH_2$) in an electron transport chain is then used to make ATP.

Like glycolysis, many of the Krebs cycle's intermediates are entry points for amino acid, nucleotide and lipid catabolism, as well as a source of carbon skeletons for synthesis (anabolism) of the same compounds. These pathways are shown in purple and green, respectively. *Details have been omitted from these other pathways. Single arrows may represent several reactions and other carbon compounds not illustrated may be required to complete a particular reaction.*

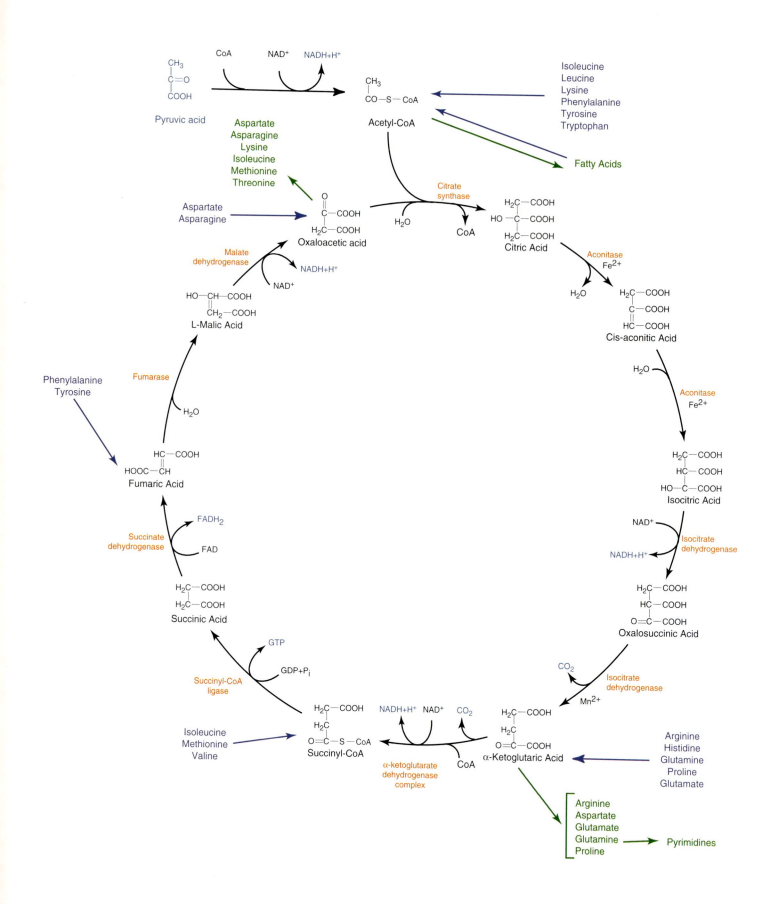

**Figure C-1.** The Krebs Cycle and associated pathways.

*A Photographic Atlas for the Microbiology Laboratory*

# Commercial Microscope Slides Used in Preparation of the Atlas

Most photomicrographs of specimens routinely available in introductory microbiology classes were taken of slides prepared by the authors. However, commercial microscope slides were used for photomicrographs of pathogens and techniques not routinely done by students. Following is a list of the commercial slides used.

## Carolina Biological Supply

*2700 York Rd.*
*Burlington, NC 27215*

| ATLAS FIGURE | SUBJECT | CATALOG NUMBER |
|---|---|---|
| 3-10 | *Vibrio fischerii* | Ba 145 |
| 4-10 | *Clostridium tetani* - spores | Ba 60 |
| 4-13 | *Spirillum volutans* - flagella | Ba 017 |
| 4-15 | *Proteus vulgaris* - flagella | Ba 017a |
| 4-17 | *Bacillus cereus* - nucleoplasm | Ba 016N |
| 4-18 | *Bacillus cereus* - PHB granules | Ba 014 |
| 8-5 | Eosinophil | F6 H6456 |
| 8-6 | Basophil | PH 1092 |
| 8-11 to 8-13 | Lymph Node | H 9150 |
| 8-16 | Lung | H 2460 |
| 8-18 | Spleen | H 9150 |
| 9-10 | *Entamoeba histolytica* - cyst | Z 115 |
| 9-12 | *Balantidium coli* | Z 465 |
| 9-19 | *Trypanosoma gambiense* | Z 260 |
| 9-28 | *Saccharomyces* Ascospore | B 261b |
| 9-32 | Rhizopus *Sporangium* | B 224 |
| 9-33 to 9-36 | *Rhizopus Gametangia* | B 225 |
| 9-39 & 9-40 | *Penicillium* | B 252 |

## The Biology Store (formerly College Biological Supply)

*P.O. Box 2691*
*Escondido, CA 92033*

| ATLAS FIGURE | SUBJECT | CATALOG NUMBER |
|---|---|---|
| 9-8 | *Paramecium* - fission | Z1.313 |

## Triarch

*Dept. SW 51 - Box 98*
*Ripon, WI 54971-0098*

| ATLAS FIGURE | SUBJECT | CATALOG NUMBER |
|---|---|---|
| 3-19 & 3-20 | *Corynebacterium diphtheriae* | 4-66 |
| 4-11 | *Corynebacterium diphtheriae* | 4-66 |
| 4-12 | *Pseudomonas aeruginosa* - flagella | 4-106a |
| 8-10 | Kupffer Cell - Liver | HK 10-114 |

## Turtox Microscope Slides

(Note: These slides were manufactured prior to Ward's purchase of Turtox. Please contact Ward's to determine availability and corresponding catalog numbers.)

| ATLAS FIGURE | SUBJECT | CATALOG NUMBER |
|---|---|---|
| 3-9 | *Treponema pallidum* | BC 4.81 |
| 4-6 | *Klebsiella pneumoniae* - capsule | BC 4.9 |
| 8-9 | Mast Cell - LAT | H2.13 |
| 9-5 | *Amoeba* | Z1.11 |
| 9-6 | *Euglena* | B1.211 |
| 9-7 | *Paramecium bursaria* | Z1.3146 |
| 9-18 | *Trypanosoma cruzi* | P1.238 |
| 9-21 to 9-24 | *Plasmodium falciparum* - ring stage | P1.4625 |
| 9-41 | *Penicillium* Conidia | B2.52 |
| 9-42 | *Aspergillus* Conidia | B2.531 |

## Ward's Natural Science Establishment

*P.O Box 92912*
*Rochester, NY 14692-9012*

| ATLAS FIGURE | SUBJECT | CATALOG NUMBER |
|---|---|---|
| 3-13 | *Neisseria gonorrhoeae* | 90 W 2060 |
| 3-16 | *Sarcina maxima* | 90 W 0553 |
| 4-9 | *Clostridium botulinum* - spores | 90 W 2029 |
| 4-14 | "*Pseudomonas reptilivora*" - flagella | 90 W 7573 |
| 8-1 to 8-4 | Segmented Neutrophil | 93 W 6540 |
| 8-7 & 8-8 | Monocyte | 93 W 6540 |
| 8-14 | Thymus | 93 W 4122 |
| 8-15 | Tonsil | 93 W 6555 |
| 8-17 | Peyer's Patch | 93 W 4102 |
| 9-9 | *Entamoeba histolytica* - trophozoite | 92 W 4083 |
| 9-14 | *Giardia lamblia* - trophozoite | 92 W 4235 |
| 9-16 | *Trichomonas vaginalis* - trophozoite | 92 W 4273 |
| 9-17 | *Leishmania donovani* - promastigotes | 92 W 4255 |
| 9-25 | *Toxoplasma gondii* | 92 W 4836 |
| 9-43 | *Candida albicans* | 90 W 7552 |
| 9-44 & 9-45 | *Pneumocystis carinii* infected lung | 92 W 4857 |

Battone, Edward (Editor). *Schneierson's Atlas of Diagnostic Microbiology.* Abbott Laboratories, 1984.

Braude, Abraham I., Charles E. Davis and Joshua Fierer (Editors). *Infectious Diseases and Medical Microbiology.* W.B. Saunders Co., 1986.

Brooks, Geo. F., Janet S. Butel and L. Nicholas Ornston. *Jawetz, Melnick & Adelberg's Medical Microbiology, 20th Ed.* Appleton & Lange, 1995.

Brown, Barbara A. *Hematology - Principles and Procedures, 6th Ed.* Lea and Febiger, 1993.

DIFCO. *DIFCO Manual–Dehydrated Culture Media and Reagents for Microbiology, 10th Ed.* DIFCO Laboratories, 1984.

Gerhardt, Philipp (Editor-in-Chief). *Manual of Methods for General Bacteriology.* American Society for Microbiology, 1981.

Gillespie, Stephen. *Medical Microbiology Illustrated.* Butterworth-Heineman Ltd., 1994.

Holt, John G. (Editor). *Bergey's Manual of Determinative Bacteriology, 9th Ed.* Williams and Wilkins, 1994.

Howard, Barbara J. *Clinical and Pathogenic Microbiology, 2nd Ed.* Mosby-Yearbook Inc., 1994.

Junqueira, L. Carlos, Jose Carneiro and Robert O. Kelley. *Basic Histology, 8th Ed.* Appleton & Lange, 1995.

Koneman, Elmer W., *et. al. Color Atlas and Textbook of Diagnostic Microbiology,* 4th Ed. J.B. Lippincott Company, 1992.

Krieg, Noel R. and John G. Holt (Editor-in-Chief). *Bergey's Manual of Systematic Bacteriology, Volume 1.* Williams & Wilkins, 1984.

Laskin, Allen I. and Hubert A. Lechevalier. *CRC Handbook of Microbiology, Volume V, 2nd Ed.* Chemical Rubber Company, 1984.

Lehninger, Albert L., David L. Nelson, and Michael M. Cox. *Principles of Biochemistry, 2nd. Ed.* Worth Publishers, 1993.

Lennette, Edwin H. (Editor-in-Chief). *Manual of Clinical Microbiology, 4th Ed.* American Society for Microbiology, 1985.

MacFaddin, Jean F. *Biochemical Tests for Identification of Medical Bacteria, 2nd Ed.* Williams & Wilkins, 1980.

Mandlestam, Joel, Kenneth McQuillen and Ian Dawes. *Biochemistry of Bacterial Growth, 3rd Ed.* Blackwell Scientific Publications, 1982.

Moat, Albert G. and John W. Foster. *Microbial Physiology, 3rd Ed.* Wiley-Liss, Inc., 1995.

Murray, Robert K., Daryl K. Granner, Peter A. Mayes and Victor W. Rodwell. *Harper's Biochemistry, 23rd Ed.* Appleton & Lange, 1993.

Roitt, Ivan, Jonathan Brostoff and David Male. *Immunology, 3rd Ed.* Mosby-Year Book Europe, Limited, 1993.

Ryan, Kenneth J. (Editor). *Sherris Medical Microbiology - An Introduction to Infectious Diseases, 3rd Ed.* Appleton and Lange, 1994.

Sneath, Peter H. A., Nicholas S. Mair, M. Elisabeth Sharpe and John G. Holt (Editor-in-Chief). *Bergey's Manual of Systematic Bacteriology, Volume 2.* Williams & Wilkins, 1986.

Stine, Gerald J. *AIDS Update 1994-1995.* Prentice-Hall, Inc. 1995.

Stites, Daniel P., Abba I. Terr and Tristram G. Parslow. *Basic and Clinical Immunology, 8th Ed.* Appleton & Lange, 1995.

Stokes, E. Joan and G.L. Ridgeway. *Clinical Microbiology, 7th Ed.* Edward Arnold (UK) Williams and Wilkins, 1988.

Voet, Donald and Judith Voet. *Biochemistry, 3rd Ed.* John Wiley & Sons. 1995.